Andréa Bicca Noguez Martins et al

From planting to harvest: an agronomic review

Andréa Bicca Noguez Martins et al

From planting to harvest: an agronomic review

Literature review

ScienciaScripts

Imprint
Any brand names and product names mentioned in this book are subject to trademark, brand or patent protection and are trademarks or registered trademarks of their respective holders. The use of brand names, product names, common names, trade names, product descriptions etc. even without a particular marking in this work is in no way to be construed to mean that such names may be regarded as unrestricted in respect of trademark and brand protection legislation and could thus be used by anyone.

Cover image: www.ingimage.com

This book is a translation from the original published under ISBN 978-620-6-76157-0.

Publisher:
Sciencia Scripts
is a trademark of
Dodo Books Indian Ocean Ltd. and OmniScriptum S.R.L publishing group

120 High Road, East Finchley, London, N2 9ED, United Kingdom
Str. Armeneasca 28/1, office 1, Chisinau MD-2012, Republic of Moldova, Europe
Printed at: see last page
ISBN: 978-620-7-96061-3

ORGANISERS ANDRÉA BICCA NOGUEZ MARTINS JOICE FERNANDA LÜBKE BONOW

PREFACE

In order to elucidate the different aspects of planting to harvest, it is necessary to constantly update technologies.

Thus, a group of teachers and students from the Federal Institute of Education, Science and Technology of Rio Grande do Sul (Instituto Federal de Educação, Ciência e Tecnologia Sul-rio-grandense)
Campus Bagé have joined forces to compile existing information on seed analyses. It is hoped that this publication will contribute to encouraging studies on this subject.
We wish you all a great read!

AUTHORS/EDITORS

Amanda Sanches

Andréa Bicca Noguez Martins

Andrei Pigatto Daniela Cunha

Joice Fernanda Lübke Bonow

Kailaine Reetez Kanauã Castro Kauã da Rosa

Leandro Cavalheiro Luid Albanio

Luis Eduardo Souza Maicon Silva

Maria Eduarda Marques Maria Eduarda S. Dias

Thainara Oliveira Thuany Pereira Viviane Trassante

SUMMARY

CHAPTER 1

DEPTH OF SOWING

Luis Eduardo Souza Maria Eduarda S. Dias
Joice Fernanda Lübke Bonow Andréa Bicca Noguez Martins

Proper soil preparation is extremely important when growing soya. This preparation is carried out to provide satisfactory conditions for sowing, germination, seedling emergence, plant development and yield. Soil compaction on the sowing row is of great importance because it causes changes inside the soil, modifying a large part of the physical environment in which the crop grows.

The four physical soil factors that need to be considered when assessing crop responses to a particular type of tillage are: moisture, temperature, aeration and mechanical resistance of the soil to penetration, and soil compaction can influence all of these, affecting the plant during its development cycle (SANTOS, 2005). The sowing depth should be between 3 and 5 centimetres. The fertiliser needs to be distributed next to and below the seed, as direct contact affects the amount and absorption of water.

Defining the seeding depth of a crop is extremely important to guarantee an adequate final stand and, consequently, successful productivity and good crop establishment (SCHMIDT et al., 1999). Numerous variables can affect sowing quality, with sowing speed being one of the most important (KURACHI et al., 2006).

Mantovani et al. (1999) observed that increasing the speed at which the tractor and seeder move changes the peripheral speed of the drilled disc, causing mechanical damage to the seeds and compromising the occupation of the cells and clamping fingers. Studies carried out by Silva, Kluthcouski and Silveira (2000) showed that the uniformity of the spacing between the maize seeds in the sowing line is also affected by the travelling speed, and was considered excellent at a speed of 3.0 km h-1, fair at 6.0 and 9.0 km h-1 and unsatisfactory at 11.2 km h-1.Another factor that affects crop development, related to the sowing process, is the depth at which the seed is deposited. Normally, maize is sown between 3 and 7 cm deep, with an average depth of 5 cm. If these seeds are sown at a greater depth, they may cause damage to emergence, with repercussions on the initial vigour of the crop (EMPRESA..., 1996).Table 1 shows the average values for the IVE variable as a function of the different seed deposition depths and sowing speeds. It was observed that there was no significant difference between the treatments, with no difference in depth or sowing speed, corroborating studies carried out by Vian (2012). During the evaluation of the IVE (16 days) the rainfall distribution was normal for the period, with no periods of drought observed.

The interaction between sowing depth and sowing speed was also not significant. Table 1 shows that, although not significant, IVE showed a downward trend as sowing speed increased. The opposite can be seen for the different depths, which, although not significant, showed an increase in IVE.According to Ortolani, Banzato and Bortoli (1986), the best response for the speed of seedling emergence index at greater sowing depths is possibly due to the effect of soil temperature, which in turn conditions the favourable thermal environment for the initial establishment of the plant.Silva et al. (2008) point out that the depth at which seeds are deposited can affect their germination, depending on temperature, water content, the peculiarity of the seed, the physical and chemical properties of the soil, climate and crop management, among other factors.

Koakoski et al. (2007) and Weirich et al. (2007) state that the greater the deposition depth, the greater the energy consumption during emergence, as well as the damage caused by low temperatures and low oxygen levels; the shallower the depth, the greater the susceptibility of the seed to water stress. Fornasiere Filho (1992) points out that deep sowing (above 7 cm) can delay seedling emergence and, in certain cases, prevent it altogether, due to the inability of the seedlings to elongate until they reach the light.

Table 1: Mean test for the variables emergence speed index (ESI) as a function of different sowing depths and speeds (Mamborê-PR, 2012).

Profundidade (m)	
0,03	10,68 a
0,05	11,94 a
Velocidade (km h^{-1})	
3	11,89 a
6	11,03 a
9	11,01 a

Means followed by the same letter in the column do not differ statistically from each other using the Tukey test at the 5% probability level.

Evaluating the variables flawed plants (FP); double plants (DP) and normal spacing (EN), the average values of which are shown in Table 2, it was observed that there was no significant difference between the sowing depths, but among the speeds studied, the speed of 9 km h-1 was agronomically inferior.

Table 2 shows that the depth of seed deposition did not significantly influence the variables studied. The speeds of 3 km h-1 and 6 km h-1 showed the best distribution for normal plant spacing. An inferior result was observed for the travelling speed equivalent to 9 km h-1.

Table 2: Test of means for the variables flawed plants (PF); double plants (PD); and normal spacing (EN), as a function of different depths and sowing speed (Mamborê-PR, 2012).

Profundidade (m)	PF	PD	EN
0,03	12,41 a	6,25 a	14,0 a
0,05	11,83 a	6,83 a	13,3 a
Velocidade (km h^{-1})	PF	PD	EN
3	12,12 a	6,37 a b	16,37 a
6	11,12 a	5,12 b	16,12 a
9	13,12 a	8,12 a	8,50 b
C.V. (%)	19,56	35,53	34,79

Means followed by the same letter in the column do not differ statistically from each other using the Tukey test at the 5% probability level. Soya cultivation is among the agricultural activities with the greatest impact on world agribusiness. In Brazil, it is present in the majority of areas destined for grain production and has been the main reason for the great expansion of the activity throughout the country.

Agricultural activities in general are strongly subject to the effects of weather and climatic conditions (FIORIN and ROSS, 2015). Climatic factors and elements directly interfere with farmers' decision-making, influencing the choice, for example, of the ideal sowing depth and the cultivars best suited to the climatic conditions of the new growing regions. The use of sowing depth in conjunction with seeds inoculated with rhizobacteria has already provided benefits for the Poaceae family, in species such as wheat (ZOZ et al., 2019), rice (OLIVEIRA et al., 2020) and forage plants, especially in conditions of biotic stress (LOREDO-OSTI et al., 2004).In soya cultivation, studies have shown positive effects of inoculation with Bradyrhizobium and Azospirillum in relation to emergence and initial development (BOSCHETTI and SIMONETTI, 2018), as well as increases in productivity, grain protein content, seed physiological quality, nodulation, tolerance to adverse conditions, among others (SOUZA et al., 2020; ZEFFA et al., 2020). The aim of this study was to assess the effect of inoculation with Bradyrhizobium japonicum applied alone and together with Azospirillum brasilense on two soya cultivars, sown at different sowing depths.

MATERIALS AND METHODS

Two experiments were carried out on a rural property located in the municipality of Couto Magalhães in the state of Tocantins (coordinates 8°20' S and 49°11' W, and average altitude of 195m), in the 2019/2020 agricultural year. The seeds were sown on 13 December 2019 and 13 January 2020, respectively, in polyethylene pots under field conditions. Data on minimum and maximum temperature and total rainfall during the work were collected at the experimental site using an INCOTERM rain gauge. The soil used in the experiments belongs to the Plinthosol class, collected in an agricultural area with a history of soybean cultivation for at least 10 consecutive years, whose physical-chemical characterisation was 16; 20; 66; 63.34% clay, silt and sand, respectively; pH in CaCl2: 6.22; Ca: 3.05 cmolc.Dm^{-3} ; Mg 1.19 cmolc dm^{-3} ; Al: 0.0 cmolc dm^{-3} ; Al +H: 1.90 cmolc dm^{-3} ; K: 0.42 cmolc dm^{-3} ; CTC: 6.56 cmolc dm^{-3} ; P-mel: 42.94 mg dm^{-3} ; S: 3.23 mg dm^{-3} ; Cu: 0.20 mg dm- 3; Zn: 1.65 mgdm^{-3} ; Fe:

80.25 mg dm^{-3} ; Co: 0.06 mg dm^{-3} ; B: 0.25 mg cm^{-3} ; Mo: 0.11 mg dm^{-3} ; Mn: 7.59 mg dm^{-3} ; M.O.:1.48%, V%:71.04

The experimental design used in each experiment was entirely randomised with 20 treatments and 4 replications. The treatments were arranged in a 2 x 5 x 2 factorial design, represented by two cultivars (Bonus 8579 IPRO and NS 8383 RR), five sowing depths (2, 4, 6, 8 and 10 cm) and two seed inoculation treatments (presence and absence of Azospirillum brasilense).

The inoculation methods were; a) inoculation with 600 mL 50 kg^{-1} of Bradyrhizobium japonicum (strains Semeia 5079 and Semeia 5080 - 5.0 x 10 9 viable cells mL^{-1}); b) co-inoculation 600 mL 50 kg^{-1} of Bradyrhizobium japonicum + 400 mL 50 kg^{-1} of Azospirillum brasilense (strains AbV5 and AbV6 - 2.0 x 10 8 viable cells mL^{-1}). Both strains were obtained from commercial products registered with the Ministry of Agriculture, Livestock and Supply.

On the day of sowing, the seeds were inoculated with B. japonicum and/or co-inoculated with B. japonicum + A. brasilense. Coinoculation was achieved by combining the two bacteria and applying them homogeneously to the seeds, firstly B. japonicum, and then A. brasilense, at the dosages described above. The seeds used in the trials were purchased with seed treatment carried out by the companies themselves. The soil was corrected so that nutrients were not a limiting factor in the soil, using the dosage to obtain maximum productivity.After sowing, the number of emerged seedlings was counted daily until it was constant. Seedlings were considered to have emerged when they broke through the soil and could be seen with the naked eye from any angle. Based on the number of emerged seedlings, the emergence percentage (LABOURIAU and VALADARES, 1976) and the emergence speed index (MAGUIRE, 1962) were determined.

In conclusion, the triple interaction Season x Cultivar x Depth of sowing (E x C x PF) for the speed index (IV) trait shows that the cultivars behave differently depending on the environmental conditions arising from the seasons and depths of sowing, and the results are broken down. In addition, the fluctuations in maximum temperatures observed in the first season, as well as affecting the germination process of the cultivars' seeds, may also have affected the symbiosis process.The increase in depth had a negative influence on emergence characteristics, regardless of inoculation management and sowing time.

REFERENCES

BOSCHETTI, E.L; SIMONETTI, A.P.M.M. Influence of bradyrhizobium and azospirillum co-cultivation on the initial development of soya beans. revista cultivando o saber, special edition, p. 44-52, 2018.

FIORIN, T.T.; ROSS, M.D. Climatologia agrícola. Santa Maria: Federal University of Santa Maria, Polytechnic College; e-tec Brazil Network, 2015. 82p.

FORNASIERE FILHO, D. Maize cultivation. Jaboticabal: FUNEP, 1992. 272 P.

KURACHI, S. A. H. ET AL. Technological evaluation of seeders and/or fertilisers: treatment of test data influence of travel speed on maize sowing. Engenharia Agrícola, Jaboticabal, v. 26, n. 2, p. 520-527, 2006.

LABOURIAU, L.G.; VALADARES, M.E.B. On the germination of seeds calotropis procere (ait.) Ait. F. Anais da academia brasileira de ciências, v.48, n.2, p.263-284, 1976.

MANGIRE, J.D. speed of germination-aid in relation evaluation for seedling emergence vigour. Crop science, v.2, n.2, p.176-177, 1962.

ORTOLANI, A. F; BANZATO, D. A; BORTOLI, N. M. Influence of soil depth and compaction on the emergence and development of grain sorghum (Sorghum Bicolor l. Moench). In: Brazilian Congress of Agricultural Engineering, 15, 1986, Botucatu. Proceedings... Botucatu: Brazilian Society of Agricultural Engineering, 1986. p. 24-32.

PLAZINSKI, J.; ROLFE, B.G. Influence of azospirillum strains on the nodulation of clovers by rhizobiology strains. Applied and environmental microbiology, v.49, n.4, p.984-989, 1985.

SCHMIDT, A. V. ET AL. Fertiliser seeder for no-till. Porto Alegre: Emater, 1999. 56 P.

SOUZA, F.G.DE; SILVA, E.L.S. DA; ALVAREZ, R. DE C. F.; ZANELLA, M. S.;

LIMA, S. F. DE. Inoculation and co-inoculation of Bradyrhizobium Japonicum and azospirillum brasilense in soya cultivation. Research, Society and Development, v.9, n.6, e170963553, 2020.

WEIRICH NETO, P. H. ET AL. Depth of deposition of maize seed in the campos gerais region, Paraná. Engenharia agrícola, Jaboticabal, v. 27, n. 3, p. 782- 786, 2007. http://dx.doi.org/10.1590/S0100-69162007000400022.

ZEFFA, D.M.; FANTIN, L.H.; KOLTUN, A .; OLIVEIRA, A.L.M.DE; NUNES,

M.P.B.A.; CANTERI, M.G.; GONÇALVES, L.S.A. effects of plant growth- promoting rhizobacteria on co-inculation with bradyrhizobium in soybean crop: a meta-analysis of studies from 1987 to 2018. Peerj, v.8, e7905, p. 1-19, 2020.

ZOZ, T.; VENDRUSCOLO, E.P.; WITT, T.W.; OLIVEIRA, C.E.DA. S.; ZOZ, J.; ZOZ,

A. Does Azospirillum brasilense and stimulate to improve the initial growth of wheat sown at greater dephths, Revista brasileira de ciências agrárias, v.14, n.1, e5604, 2019.

CHAPTER 2

WEEDS IN RICE, SOYA AND WHEAT CROPS

Kailaine Reetez Kanauã Castro
Maria Eduarda Marques Joice Fernanda Lübke Bonow Andréa Bicca Noguez Martins

• RICE

1 - GENERAL INTRODUCTION

Rice is one of the world's most important crops. It is the main nutritional source for the population of developing countries. It comes mainly from irrigated rice fields in Rio Grande do Sul and Santa Catarina. These states contribute around 60 per cent of national production (SOSBAI, 2019). However, like any other crop, it is susceptible to invasive plants and the damage caused by them depends on various factors, such as the cultivar used, soil fertility and fertiliser used, crop population and spacing, weed species and their population. In areas where there is a large infestation of weeds, the damage caused can amount to a total loss of production (Cobucci, 1998).

2 - WEEDS MOST COMMONLY FOUND IN IRRIGATED RICE CROPS

2.1 - WEEDY RICE:

Figure 1 and 2: Fields contaminated with weedy rice.

Source: https://www.epagri.sc.gov.br/wp-content/uploads/2020/03/arroz-daninho.jpg
(Accessed: 21 July 2024)

It belongs to the same species as cultivated rice (oryza sativa L.), which makes it more difficult to manage. One of its main adaptive strategies is the degrading of its seeds before the cultivated rice is harvested. This phenomenon allows the new generation of "shoots" of the weed to reach the seed bank, contributing to its spread (Andres, 2021). Still according to the author, without being able to be harvested simultaneously with the crop and consequently removed from the environment. These seeds separate from the mother plant with varying degrees of dormancy, which ensures that they emerge at different times. distribution of infestation over time. In addition to the aforementioned evolutionary adaptations, weedy rice seeds may have greater longevity than those of cultivated rice and the ability to emerge in a wide range of environmental conditions, especially when incorporated at depths greater than 10 cm;

2.2 - THE GENUS ECHINOCHLOA, COMMONLY KNOWN AS RICE GRASS

Figures 3, 4 and 5: Rice fields contaminated with rice grass.

Figure 1: https://dynamicassets.basf.com/is/image/basf/CapimArroz.3?dpr=off&fmt=webp-alpha&fit=crop%2C1&wid=1280&hei=960 (Accessed: 21 July 2024)

Figure 2: https://www.agroseguro.cnptia.embrapa.br/templates/capim_arroz.html (Accessed: 21 July 2024)

Figure 3: https://www.agrolink.com.br/agrolinkfito/manejo-integrado/manejo-integrado-de-weeds/integrated-management-of-weeds-in-rice_483564.html (Accessed: 21 July 2024)

It has several species that cause severe damage to agriculture worldwide, infesting more than 36 crops in 61 countries (Norris et al., 2001). The species is a problem because of its tolerance to soil flooding, its ability to interfere with crop growth and because it is widespread throughout the rice-growing area of Rio Grande do Sul. As it has a C4 metabolism, it is more efficient at utilising environmental resources than cultivated rice, producing large quantities of seeds in a short space of time, even in conditions that are adverse for most species. Other characteristics that favour rice grass are high prolificacy, different levels of dormancy and the ability to germinate in a wide range of favourable conditions, resulting in large infestations during the establishment of irrigated rice and the start of irrigation (Machado et al., 2023). The reduction in grain yield varies according to the season and the efficiency of the weed control used, which can be higher than 80 per cent due to rice grass interference (ANDRES and MENEZES, 1997; MENEZES et al., 1999);

2.3 Angiquinho (genus Aeschynomene sp.) Belongs to the fabaceae group and has a total of approximately 160 species worldwide, of which 84 species are distributed throughout the American continent, with an absolute predominance in the Neotropics, with 52 species occurring in Brazil (Fernandes, 1996);

Figure 6: Adult angiquinho plants (Aeschynomene denticulata and Aeschynomene indica) in the irrigated rice farming.

Because it is a vigorous plant, even when present in low infestations, it is very damaging to the rice crop, because as well as competing for production factors, it makes mechanical harvesting very difficult. The dark-coloured seeds, when mixed with the rice, depreciate the quality and consequently the commercial value of the product (Andrade, 1986). Aeschynomene species are of little positive economic importance, contributing only to the fixation of nitrogen in the soil, a common occurrence among fabaceous species (Kissmann & Groth, 2000), thanks to the development of nodules symbiotically associated with nitrifying bacteria.

2.4 Leather hat (Sagittaria montevidensis) stands out among the weeds that occur in irrigated rice crops because it is a rhizomatous aquatic species with characteristics of a competing plant and adaptation to a wide ecological range, which allows it to inhabit flooded environments with different physical and chemical characteristics (Cassol et al., 2008). High levels of infestation of this species have been observed due to difficulties in controlling it, allowing it to enrich the seed bank, which leads to an increase in its occurrence to levels that affect crop productivity (Gibson et al., 2001).

Figure 7: Leather hat

3 - CONTROL METHODS

The weed control strategy should associate the best method with the right time, before the critical period of competition. The choice of method, or combination of methods, must be related to local labour and implement conditions, taking into account cost analysis (Mascarenhas et al., 2007).

3.1 PREVENTIVE

The preventive method aims to prevent the introduction, establishment and spread of species in areas that are not yet infested. To this end, one of the main actions is to use seeds with a high degree of purity. National legislation sets limits for seeds of tolerated weed species and determines which species are prohibited in commercial seed lots. This prevents the use of seeds with weed propagules, especially those that are difficult to control, in areas that are not yet infested. Other precautions are necessary, such as avoiding the use of manure, straw or compost containing weed propagules, thoroughly cleaning agricultural equipment before it enters the field or after it has been used in plots where there are problem species and controlling these plants near the edges of the tracks (Mascarenhas, 2007).

3.2 MECHANICAL CONTROL:

Use hand or tractor implements to remove weeds. Soil preparation with different equipment before sowing or in the off-season is an efficient alternative for mechanically controlling plants established weeds and to reduce the seed bank in the soil (MALARDO, 2017). Preparing the soil with a plough harrow favours the germination and proliferation of weeds compared to deep ploughing, after incorporating the cultural remains with a plough harrow, which places most of the weed seeds at a depth of approximately 30cm, hindering emergence and favouring rotting (Machado, 2023);

3.2 CULTURAL CONTROL

Irrigated rice uses a water layer in its cultivation, which is an important tool for physical weed control. The water layer reduces oxygen in the soil, preventing the seeds of various species from germinating (Machado, 2023). Providing rice plants with the capacity to manifest their maximum productive potential and compete with weeds (use of cultivars adapted to the climate and soil; good quality seeds; balanced fertilisation; appropriate sowing time; spatial arrangement of plants; pest and disease management; crop rotation). (Cobocci, 2001). In addition, crop rotation of rice with crops such as soya and maize is also becoming a reality, as long as the drainage of the area is adequate and subsequently the system of semi-direct sowing of rice on the remnants of the previous crop is adopted, with soya being the most common example, in which the only soil disturbance practice is the construction of ditches and the desiccation of succession vegetation. The introduction of these new species into the irrigated rice growing environment also allows for the use of herbicides with alternative mechanisms of action, which can control species that have developed resistance to the herbicides used in rice cultivation. The main benefit of this control method is the reduction of the weed seed bank in the soil (Andres, 2021).

3.3 CHEMICAL CONTROL

Pre-sowing: these are applications carried out to implement the no-till and minimum tillage systems. In pre-sowing, weeds and/or green winter cover are eliminated before the rice is sown. This is a key operation, as it replaces tillage operations in the elimination of weeds and the formation of mulch. This phase is called management or desiccation, when non-selective, all-action herbicides are used. Pre-emergence: the application is made shortly after sowing and before the weeds and rice emerge. For the herbicides to perform well, it is important that the soil is damp or that there is rain to incorporate the herbicide into the topsoil.

Post-emergence: the application is made after the crop and weeds have emerged. The herbicides should be applied when the weeds are at the initial stage of development. At this stage, the weeds are not yet competing with the rice crop. Post-emergence after flooding (benzedura): This consists of applying the herbicide over the water surface in post-emergence. It can be applied by aeroplane or by the benzedura method. In aerial applications, late application is not recommended when the weeds have more than two tillers, as the crop and weeds make it difficult for the product to come into contact with the water (Noldin, 2021).

- **SOYBEANS**

1 - GENERAL INTRODUCTION:

The cultivation of soya (Glycine max (L.) Merrill) stands out as one of the most significant oilseed species in the global economy, playing a vital role in various production chains (CAMPEÃO et al., 2020). Its role has expanded considerably in recent years. Widely used by industry, soya is used in the production of vegetable oil and in the manufacture of animal feed. It is also gaining prominence as an alternative source of biofuel, reflecting the growing importance of soya not only as an agricultural crop, but also as a strategic resource for the production of renewable energy (ANdRADE NETO; RAIHER, 2023).

2 - WEEDS MOST COMMONLY FOUND IN SOYA CROPS

The presence of weeds in soya crops can affect the development of the crop by promoting competition for environmental resources such as water, light and nutrients, reducing the availability of these resources for the crop and causing a reduction in grain yield due to the effects of interference on the variables that define crop yield. The time when weed control begins has a major influence on plant growth and grain yield in soya beans (Rizzardi et al., 2004). The negative effects of the weed community on crops result from both the increase in weed density and the length of the interference period (GHERSA & HOLT, 1995). With regard to density, Vandevender et al. (1997) observed that an increase in weed density reduced yields in irrigated rice. Rizzardi et al. (2003) found that the degree of interference exerted by dicotyledonous weeds on soya crops depends on the weed species present and its density.

In practice, the effects of interference are irreversible, and there is no recovery of development or productivity once the stress caused by the presence of weeds has been removed (KOSLOWSKI et al., 2002). The effects of weed interference on crop characteristics can jeopardise the development of reproductive structures and affect grain yield components

(LAMEGO et al., 2004). According to Board et al. (1995), in soya beans, the number of pods is the characteristic most responsive to changes caused by the stress of competition from competing species, while the number of grains per pod and average grain weight have greater individual control, showing little range of variation due to the environment.

3 – MAINFAMILIES ACCORDING ACCORDING TO O MANUAL FOR THE IDENTIFICATION OF WEEDS IN SOYA CULTIVATION

3.1 AMARANTHACEAE FAMILY

The Amaranthaceae family has two genera that are very important in soya cultivation, Alternanthera and Amaranthus. They are usually plants with a high production of small seeds that are easily dispersed. Some species, such as Amaranthus retroflexus, have a C4 photosynthetic metabolism, which gives them a high competitive capacity in periods of good sunshine and restricted rainfall. Some species are important not only for reducing soya yields, but also for interfering with mechanical harvesting and for being an intermediate host for nematodes. The relative importance of amaranth tends to increase as the pH and fertility of the soil rises. The main species that occur in soya cultivation are listed below.

Figure 8: Alternanthera tenella

Common name: Fire extinguisher

Figure 9: Amaranthus deflexus

Common name: Shrew

3.2 ASTERACEAE FAMILY (COMPOSITAE)

The Asteraceae family has several species that are characterised as important weeds in soya cultivation. The representatives of this family show great variability in their growth forms, from prostrate plants with little biomass, such as the creeping tick, to erect plants of great size, such as the buva and the maria-mole, which create problems for the operation of mechanical harvesting. It harbours some species that are highly likely to show resistance to herbicides, such as the black woodpecker, others that are highly capable of producing and dispersing spreaders, such as the bullwort, and some that can grow in poorly fertile soils, such as the crabgrass. The main species that occur on soya are:

Figure 10: Acanthospermum hispidum

Common name: Carnicho de Carneiro

3.3 BRASSICACEAE FAMILY

Figure 11: Conyza spp.

Common name: Buva

The Brassicaceae family, also known as the Cruciferae, has a large number of genera (320) and an even larger number of species (3200), many of which are of great importance for human consumption, such as cabbage, turnip and mustard, among others. Two species are considered to be important weeds in soya cultivation, and will be mentioned below:

Figure 12: Coronopus didymus
Common name: Mastruço

3.4 COMMELINACEAE FAMILY

Figure 13: Raphanus raphanistrum
Common name: Turnip

The Commelinaceae family has many morphologically similar species known simply as ragweeds. The most important, Commelinna benghalensis, will be described in this publication, as will Murdannia nudiflora, initially classified as Commelina nudiflora. Other species such as Commelina diffusa, Commelina erecta and Commelina villosa also occur in Brazil and have been a cause for concern in commercial soya and maize plantations in the central and southern regions of Paraná. Although they occur less frequently, every care must be taken to prevent the spread of these species, as they are all considered difficult to control chemically, especially C. villosa, as it can withstand high doses of the herbicide glyphosate.

Figure 14: Murdannia nudiflora
Common name: Murdania, Trapoerabinha

3.5 CONVOLVULACEAE FAMILY

Figure 15: Commelina benghalensis
Common name: Trapoeraba

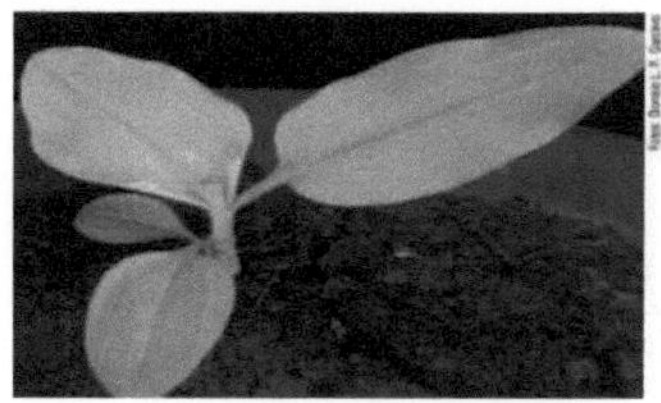

This family includes some of the most problematic weeds for soya cultivation, due to their "climbing" habit and major interference with mechanised harvesting. The most common genera are Ipomoea and Merremia and their importance tends to grow with the adoption of the direct sowing system and the cultivation of soya genetically modified for tolerance to glyphosate. Currently, the main species present in soya cultivation are those described below.

Figure 16: Ipomoea grandifolia
Common name: Corda-de-viola

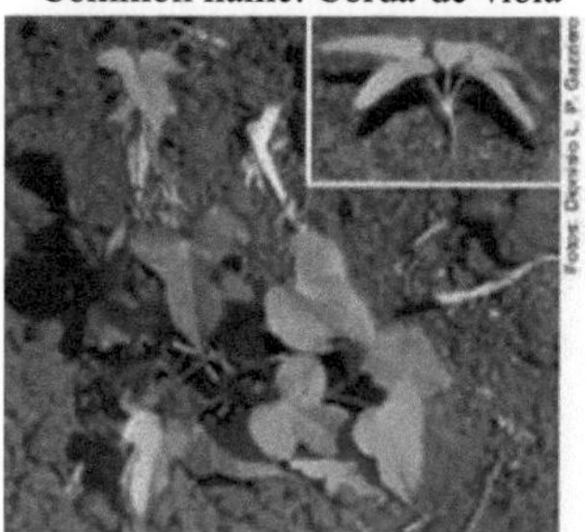

18

Figure 17: Ipomoea nil
Common name: Corda-de-viola

3.6 EUPHORBIACEAE FAMILY

The Euphorbiaceae family is home to some of the most important weeds in the soya crop, especially the peanut. The genera Euphorbia and Chamaesyce have been the most important. The latter has stood out in no-till areas, while the Euphorbia genus is home to one of the most worrying soya weeds (E. heterophylla). The main species that occur in soya crops are described here.

Figure 18: Chamaesyce hirta
Common name: St John's wort

3.7 FABACEAE FAMILY

Figure 19: Chamaesyce hyssopifolia
Common name: Swallow-wort

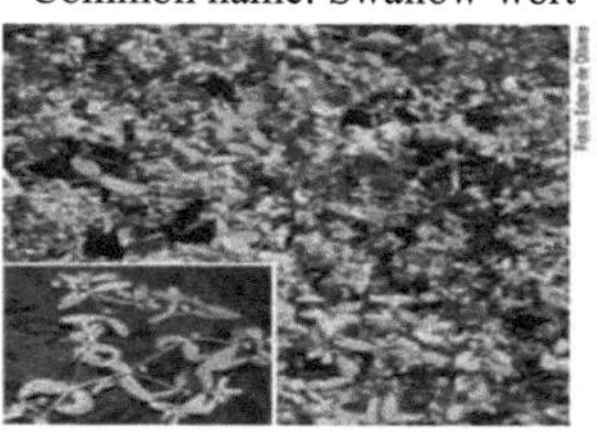

The Fabaceae family is one of the most important weeds in the soya crop. It stands out due to its taxonomic proximity to soya, and hence the similarity of its response to agricultural practices, including the application of herbicides. Several species have been selected by the continued use of some herbicides, such as fedegoso and desmodium. As well as causing damage by coexisting with the crop, these species also have a strong impact on mechanical harvesting, and are described below.

Figure 20: Desmodium tortuosum
Common name: Desmodium

3.8 LAMIACEAE FAMILY

Figure 21: Senna obtusifolia
Common name: Fedegoso

The Lamiaceae family has three species that stand out as important weeds in soya production areas. Two of them vegetate mainly during the soya off-season and another is important as a crop weed

Figure 22: Hyptys suaveolens
Common name: Cheirosa

Figure 23: Leonotis nepetifolia
Common name: Cordão-de-frade

3.9 MALVACEAE FAMILY

The Malvaceae family has numerous species that occur as weeds in the soya crop, but with very regionalised distribution, such as Wissadula subpeltata, Sida glaziovii, Sida acuta and others. Only Sida rhombifolia, which has a more widespread distribution, will be mentioned in this manual

Figure 24: Sida rhombifolia
Common name: Guanxuma, guaxuma, vassourinha

3.10 POACEAE FAMILY

The Poaceae family has numerous representatives in the soya crop and great variability in the forms of interference and control difficulties. Marmalade grass and crabgrass have a more widespread distribution in Brazil, but their importance is growing in crops in the south-east and south. Carrapicho grass is more important in the Centre-West region. Some species have a very widespread distribution due to their predominantly vegetative spread, such as the massambará grass, in contrast to the small-seeded species, such as the chickweed and the crabgrass. Others are only important due to competitive interference, while some larger plants can decisively interfere with mechanised harvesting, as is the case with colonião and massambará grasses. Certain species in this family have populations that are resistant to herbicides, such as wild oats and the crabgrass and marmalade, with biotypes registered as resistant to ACCase (Acetyl co-enzyme A carboxylase) inhibitors, or bitter crabgrass, resistant to glyphosate (EPSPs). The main species that occur in soya plantations are described below

Figure 25: Brachiaria brizantha
Common name: Brizantão

Figure 26: Brachiaria decumbens
Common name: Brachiaria grass

3.11 PORTULACACEAE FAMILY

The two important species of the Portulacaceae family in soya cultivation have different origins. Portulaca oleracea is believed to come from Asia, although some botanists believe it to be African. On the other hand, Talinum paniculatum, which also occurs in Africa, originated in America. Both are demanding in terms of soil fertility and occur more frequently in humid or irrigated areas.

Figure 27: Portulaca oleracea
Common name: purslane

Figure 28: Talinum paniculatum
Common name: Maria-gorda

3.12 RUBIACEAE FAMILY

The Rubiaceae family has three species that are considered important soya weeds. They have always been considered difficult to control in commercial crops of conventional soya; in transgenic soya resistant to glyphosate, these species have been indicated as tolerant to this molecule.

Figure 29: Richardia brasiliensis
Common name: Poaia-branca

3.13 SAPINDACEAE FAMILY

Figure 30: Spermacoce latifolia
Common name: Hot grass

The Sapindaceae family is an important species for soya cultivation, which has been selected by the herbicides used in the soya production system, gaining importance in the southern states of Brazil and in the south of Mato Grosso do Sul.

Figure 31: Cardiospermum halicacabum
Common name: Sack of Bread

3.14 SOLANACEAE FAMILY

In this family, two species stand out as important weeds in the soya crop, although others may be listed from time to time, such as the wild weed (Solanum sisymbrifolium) in the south of the country

Figure 32: Nicandra physaloides

Common name: Joá-de-capote

Figure 33: Solanum americanum

Common name: Maria-pretinha

• WEEDS IN THE WHEAT CROP.

1. INTRODUCTION:

Wheat is the most widely planted crop in the world, ranking first in terms of production volume (OLIVEIRA NETO; SANTOS, 2017). Brazil has an annual demand of 12 million tonnes and produces the equivalent of 5.80 million tonnes, less than half of its needs, importing the remaining quantity (SILVA et al., 2017). Most of the wheat (Triticum aestivum) consumed in Brazil is imported. However, the crop is the main economic winter crop in some states of the country, such as Paraná, where growers are able to produce excellent quality wheat using available technologies. One of the keys to achieving this goal is weed management (PENCKOWSKI, L.H. et al. 2003).In the world, weed interference in crops causes between 30 and 40 per cent damage to yields (PENCKOWSKI, L.H. et al. 2003). The losses caused by weed interference in wheat grain yield can be due to competition, the effect of allelopathy or, indirectly, by reducing the quality of the harvested product. Competition occurs when any factor in the environment (water, light, nutrients, etc.) is divided between the crop and the weeds, limiting the potential yield of the crop (RICE, 1984).

2. INVASIVE PLANTS THE VILLAIN OF THE CROP.

The effectiveness of a herbicide depends on various factors, such as the physicochemical characteristics and dose of the herbicide, the species to be controlled (its own structural characteristics), the stage of development and the biology of the weed (PENCKOWSKI, L.H. et al. 2003). Lorenzi (1976) estimated that crop losses caused by invasive plants are around 30 to 40 per cent and 20 to 30 per cent of world and national agricultural production, respectively. Invasive plants can reduce the quality and yield of a crop, make harvesting difficult and, in extreme cases, unfeasible. They can also increase grain moisture and drying costs, favour fermentation and increase the incidence of pests in storage (VARGAS ; ROMAN, 2005). In general, invasive plants require the same factors for their development (water, light, nutrients and physical space) as the crop, establishing a process of competition and interference that can be determined by the composition of the invasive flora (species, density and distribution), crop (species or variety, spacing and planting density), environment (soil, climate), cultural treatment (management) and period of coexistence (PITELLI, 1985; KARAM et al., 2006). If managed properly, invasive plants can increase productivity by providing food for predators, parasites and pests (SOUZA, 1991). The main weeds of wheat that will be mentioned include Buva (Conyza bonariensis), ryegrass (Lolium multiflorum), and black oats (Avena strígosa).

2.1 BUVA (Conyza bonariensis)

Figure 34: Buva - Conyza bonariensis

Source: syngenta (2018).

It is popularly known as buva and is native to South America, occurring in Uruguay, Argentina, Paraguay and the south, southeast and central-west regions of Brazil (LORENZI & MATOS, 2000). It has an annual cycle and is characterised by being very prolific, being able to produce more than 200,000 viable seeds in a single plant (BHOWMIK & BEKECH, 1993). In commercial crops, buva is considered a difficult weed to control due to its resistance to some herbicides (PYON et al., 2004). Crabgrass produces a large number of seeds, which have characteristics and structures that make them easy to disperse, characterising the species as aggressive (Kissmann & Groth, 1992). Weed control consists of adopting practices that result in the reduction of infestation, but not necessarily its complete elimination or eradication. Reducing weed interference, considering a crop, should be done up to the level where the losses due to interference are equal to the cost of control, i.e. so that they do not interfere with the economic production of the crop (SILVA et al., 1999).

2.2 BLUEBERRY (Lolium multiflorum)

Figure 35: Ryegrass (Lolium multiflorum)

Source: botanical garden.

Ryegrass is the most widely used winter fodder plant in Rio Grande do Sul, as well as in most temperate and subtropical regions of the world (Bressolin 2007). To control weed species in triticultural crops, a number of efficient and selective herbicides are available. However, their repeated use has selected resistant weed species worldwide, such as turnip resistant to herbicides that inhibit the enzyme acetolactate synthase (ALS) and ryegrass resistant to the herbicide glyphosate and ALS (Heap, 2012). Agostinetto et al. (2008) observed that the infestation of ryegrass plants led to a lower accumulation of dry biomass in the wheat crop during the coexistence period, possibly due to the lower mission and survival of the offspring.

Under conditions of competition for light, wheat plants tended to increase their height; according to Almeida and Mundstock (2001), light is a determining factor in the tillering capacity and productivity of the wheat crop. Winter cereals, when in competition for light, increase the direction of photoassimilates in the formation of stalks, in other words, there is a stioling of the plants in order to capture more light, with less investment of energy for tillering, development of leaf area and dry biomass and root growth; even grain yield can be affected because the plant invests more energy in the organ with the greatest nutritional deficit as opposed to the others (GALON et al., 2011).

2.3 OATS (Avena strigosa)

Figure 36: Oats (Avena strigosa)

Source: Wellington Bernardes

Black oats (Avena strigosa Schreb.), due to their characteristics of cycle compatibility with other species in the regional production system, high phytomass production, abundant root system and rusticity (CALEGARI et al., 1992). Despite the numerous benefits that black oats can bring to the production process, care must be taken with their management to prevent this species from becoming a weed (FONTANELI et al., 1997).

3. CONCLUSION

To put preventive control into practice, farmers should: use certified seeds, avoid moving animals from infested areas to weed-free areas, clean equipment after working in areas with undesirable weeds and control these species in canals, crop margins and paths (SILVA et al., 1999).

REFERENCES

ANDRADE NETO, Antonio Octaviano de; RAIHER, Augusta Pelinski. Socio-economic impact of soya cultivation in the minimum comparable areas of Brazil. Revista de Economia e Sociologia Rural, v. 62, p. e267567, 2023

ANDRES, A.; MARTINS, M. T. Weed management.
ANDRES, A.; THEISEN. G.; RIEFFEL. J. F.; HOFFMAN.D.; NEVES, R. competition of Rice Grass (Echinochloa crusgalli) in irrigated rice: control times and damage to Cultivar BRS Querência.

BOARD, J. E.; WIER, A. T.; BOETHEL, D. J. Source strength influence on soybean formation during early and late reproductive development. Crop. Sci., v. 35, n. 4, p. 1104-1110, 1995.

CAMPEÃO, Patrícia; SANCHES, Arthur Caldeira; MACIEL, Wilson Ravelli Elizeu. International Commodities Market: an analysis of Brazil's participation in the world soya market between 2008 and 2019. Desenvolvimento em Questão, v. 18, n. 51, p. 76-92, 2020.

CORADINI, M. C., SCHERNER, A., SCHREIBER, F., ANDRES, A., CONCENÇO, G., SILVA, J. T., CEOLIN, W. C., MOISINHO, I.S. Growth of sagittaria (Sagittaria montevidensis) as a function of water depth.

COTRIEL. Espumoso Crop Cooperative. Producers should intensify care with herbicide-resistant ryegrass. Espumoso, 2015. Available at:<http://www.cotriel.com.br/ News/producers-should-intensify-care-with-herbicide-resistant-ryegrass>.

DUURVOORT, F. E. Get to know one of the world's most aggressive plants.
FERREIRA, E. A.; CONCENÇO.G.; GALON.L.; DELGADO. M. N.; ASPIAZÚ. I.; SILVA. A.F.; FERREIRA. F. A.; MEIRA. R.M. S. Micromorphological characteristics of rice grass biotypes resistant and susceptible to quinclorac.

FERREIRA, F. B. Biology, competitive ability and genetic variability in three species of angiquinho (aeschynomene spp.) and their management in irrigated rice.

FLECK, N. G. Competition between ryegrass (Lolium multiflorum L.) and two wheat cultivars. Planta Daninha, v. 3, n. 2, p. 61-67, 1980.

FONTANELI, R.S. et al. Management of black oats as a soil cover in the no-till system. Passo Fundo: Embrapa Trigo/Projeto METAS, 1997. 18p. (METAS Project. Technical Bulletin, 2).

FONTES, J. R. A., & GONCALVES, J. R. P. (2009). Integrated weed management.

GALON, L.; AGOSTINETTO.; MORAES, P.V.D.; TIRONI, S.P.; DAL, M.T. Estimating grain yield losses in rice cultivars (Oryza sativa) due to rice grass (Echinochloa spp.) interference

GAZZIERO, Dionísio Luiz Pisa et al. Manual de identificação de plantas daninhas da cultura da soja. Embrapa Soja, 2006.

GHERSA, C. M.; HOLT, J. S. Using phenology prediction in weed management: a review. Weed Res., v. 35, n. 6, p. 461-470, 1995.

KOSLOWSKI, L. A. et al. Weed interference in the common bean crop in a direct sowing system. Planta Daninha, v. 20, n. 2, p. 213-220, 2002.

KUVA, M.A.; PITELLI, R.A.; SALGADO, T.P.; ALVES, P.L.C.A. Phytosociology of weed communities in a sugarcane-crop agroecosystem. Planta Daninha, v. 25, p.501-511, 2007.

LACERDA, M. C. Weed management.
LAMEGO, F. P. et al. Tolerance to interference from competing plants and ability to suppress by soya genotypes II. Response of yield variables. Planta Daninha, v. 22, n. 4, p. 491-498, 2004.

LORENZI, H. Weeds of Brazil: terrestrial, aquatic, parasitic and toxic. 3. ed. Nova Odessa: Instituto Plantarum, 2000. 608p.
MACHADO, A. W. Control (integrated management) of weeds in irrigated rice cultivation.

MALARDO, M. Weed management in rice cultivation

MASCARENHAS, R. E. B., COBUCCI. T. Weed control in dryland rice cultivation.

PITELLI, R.A. Weed interference in agricultural crops. Informe Agropecuário, v.11, p.16-27, 1985.

RIZZARDI, M. A. et al. Soya bean yield losses caused by interference from picão-preto and guanxuma. Ci. Rural, v. 33, n. 4, p. 621-627, 2003.

RIZZARDI, M. A.; FLECK, N. G. Methods for qualifying the foliar coverage o f weed infestation and the soya bean crop. Ci. Rural, v. 34, p. 13-18, 2004.

SILVA, A. A. da; SILVA, J. F.; FERREIRA, F.A.; FERREIRA, L. R.; SILVA, J.

F.Controle de plantas daninhas. Brasília, DF: Associação Brasileira de Educação Agrícola Superior; Viçosa, MG: Universidade Federal de Viçosa, 1999.260 p.

SILVA, A.A.; SILVA, C.S.W.; SOUZA, C.M.; SOUZA, B.A.; FAGUNDES, J.L.;

FALLEIRO, R.M.; SEDIYAMA, C.S. Phytosociological aspects of the weed community in the bean crop under different tillage systems. Planta Daninha, v.23, p.17-24, 2005.

SILVA, K. S. Angiquinho (aeschynomene denticulata).
SILVA, Tuane Araldi. Physiological performance of seeds and antioxidant metabolism of lettuce and wheat seedlings under the action of aqueous extracts of buva and ryegrass. 2015.80 f. Dissertation (Master of Science) - Graduate Programme in Seed Science and Technology, Eliseu Maciel School of Agronomy, Federal University of Pelotas, Pelotas, 2015.

VANDEVENDER, K. W.; COSTELLO, T. A.; SMITH JR., R. J. Model of rice (Oryza sativa) yield reduction as a function of weed interference. Weed Sci., v. 45, n. 2, p. 218-224, 1997.

CHAPTER 3

WATER REQUIREMENTS FOR RICE, SOYA AND WHEAT CROPS

Kauã da Rosa Luid Albanio Maicon Silva
Joice Fernanda Lübke Bonow Andréa Bicca Noguez Martins

One of the important factors that directly influences the productive development of any crop is the amount of water available in the area; a shortage or excess can hinder growth and even the formation of grains. Different levels of rainfall, temperature and humidity contribute to or interfere with the agricultural production of rice, soya and wheat crops, and having knowledge of these metrics helps not only producers to make decisions but also researchers to develop new seed and input technologies.

- **RICE**

Rice (O. sativa L.) is considered the most important cereal crop in the world, responsible for the main source of food and calories for more than half of humanity (Khush, 2005), accounting for 20% of the world population's food energy, while it is estimated that wheat provides 19% and maize 5%.

Rice is a grass native to Asia, and its grains are sold on a large scale in basically three types: brown rice, parboiled rice and polished rice. The main chemical constituents of the grain include starch, protein, lipids, ash and total fibre. Among the types of rice, brown rice is the one with the highest amounts of the above-mentioned contents: around 74% starch, 11% protein, 2.52% lipids, 1.15% ash and 11.76% total fibre; however, as the starch content decreases, the other contents increase (CONAB, 2017). In Asian countries alone, more than two billion inhabitants use rice and its derivatives as a source of 60 to 70 per cent of their daily calorie intake (Barata, 2005). The wild ancestor of O. sativa is Oryza rufipogon, widely distributed in the area between subtropical and tropical Asia, with annual, perennial and intermediate forms. There is no consensus on which of these forms is the ancestor of cultivated rice, but it certainly evolved from one of these forms through human domestication. In addition to O. sativa, the species O. glaberrima Steud. is also cultivated, which was domesticated in West Africa (Soreng et al., 2017). In Brazil, more precisely in the south of the country, it is mainly grown by flooding, and from the beginning of its cultivation in the state until the mid-1960s, productivity was no more than three tonnes. With the advancement of agricultural technology and the adoption of new cultivation and management techniques, 20 years later productivity had already reached five tonnes per hectare (CONAB, 2021). Much of this progress was mainly due to the development of new cultivars and technologies for rice farming, mainly introduced by the research institutes: the Experimental Rice Station (EEA) and the Rio Grandense Rice Institute (IRGA) (CONAB, 2017). According to data from the United Nations Food and Agriculture Foundation, Brazilian rice consumption is approximately 52.5 kilograms per inhabitant per year (husk basis). Although this is lower than the average world consumption per inhabitant (84.8 kg/inhab/year), it is considered high when compared to the per capita consumption of developed countries (16.7 kg/inhab/year). In 1999,

30

world rice production was 586.8 million tonnes, with China being the world's largest producer in 1999 with 34.2% of the total, followed by India with 21.7% of the total, the largest exporter being the United States of America with 62.5% of the total, followed by Argentina with 12.3% of the total. Asia is responsible for 90% of world production. In Latin America, Brazil stands out as the largest producer, having produced 11.4 million tonnes in the same year on an area of 3.14 million hectares (FAO, 2006).

The new rice cultivars have higher grain quality and high yield potential, as well as more upright leaf distribution, which allows them to better utilise light. This reduces self-shading, which occurs more intensely in plants with a horizontal leaf distribution, as in traditional rice cultivars. However, the new cultivars also have reduced competitive capacity with weeds, which can result in production losses (Mangravite et al., 2021).

Advances in genetic engineering are making it possible to create rice varieties that are more tolerant and adapted to different soil and climate conditions. In a study carried out by researchers at Cornell University, USA, two Escherichia coli genes (otsA and otsB genes) that synthesise trehalose were introduced into rice. This technique makes it possible to control where the introduced genes are expressed, for example, the leaf produces trehalose and the grain does not (Hussain et al., 2014). Trehalose is a disaccharide that can be found in various plants and microorganisms and is capable of protecting the structure of cells against environmental stresses such as drought and cold. Introducing these E. coli genes into rice has the potential to increase plant resistance to these types of stresses, which could be beneficial for regions where rice production is affected by long periods of drought (Singh and Singh, 2021; Hussain et al., 2014).

Irrigation of rice crops is closely related to the cultivation system adopted. The adoption of one system or another will determine differences in the start and end times of irrigation, water management and use and, above all, soil preparation. For this reason, irrigation should be planned when the crop is systematised (Nunes, 2010).In Rio Grande do Sul, rice plantations are located in six distinct regions in terms of soil, climate and land structure. Of these, at least four - the "Outer Coastal Plain", "Campanha", "Depressão Central" and "Fronteira Oeste" - have a history of water restrictions for the crop, caused by rainfall deficits combined with topographical conditions that make it difficult to accumulate water in rivers, streams and natural reservoirs (Petrini, et al., 2008).

For flooded rice in the pre-germinated system, where flooding of the area usually takes place 20 days before sowing, with drainage three days afterwards, in order to allow better primary development of the seedlings, and later, according to the water demand of the plant's phenology, this water is inserted back into the system. However, in this system there are some losses related to the amount of water applied, i.e. water can be lost through evaporation, percolation, the edges of the fields and, to a lesser extent, overflow (Marchezan et al., 2007).

- **SOYBEANS**

It's certainly one of the most widely grown cereals in the world and it's no wonder that soya is widely used on various production fronts, from oils to printing inks. Having sufficient global production to meet this demand is essential, and in order to generate quality soya, it's vital to have adequate water availability. With this in mind, researchers around the world are investing resources and time to discover and perfect the right requirements for better soya production. grain filling while preventing diseases caused by excess. Embrapa is conducting research in this area in Brazil.

Researchers, such as Costa (2001), argue that water can account for 90% of the weight of a soya plant, and also participates in many other biochemical and physical processes in the plant, such as the transport of gases, minerals and solutes, as well as acting as a thermal regulator (Costa, 2001).

The amount of water needed varies according to the cultivar planted, but in general the average can range from 450 mm/cycle to 800 mm/cycle, varying according to the plant's establishment phase, reaching a maximum ceiling after flowering and decreasing as the grains fill, as shown below. (Berlato et al., 1986; Bergamaschi et al., 1999).

Figure 1: Water requirements according to production phase.

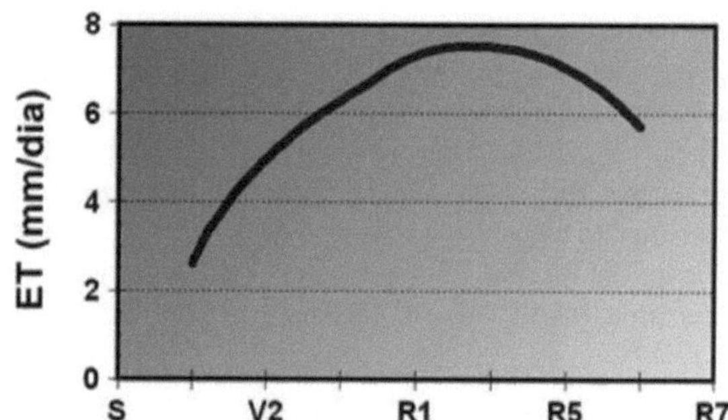

Source: Embrapa book - Soybean production technologies (adapted from Berlato et al., 1986)
(Accessed on: 10 July 2024)

Another factor that is part of the equation for soya water consumption is evapotranspiration, which varies according to the climatic conditions of the region where the crop is grown. Knowledge of total evapotranspiration (comprising water loss from both the plant and the soil) over the course of the plant's cycle is essential to help producers make decisions, both in terms of choosing the right cultivar and in terms of correct management, optimising production performance. In other words, this information also helps the producer to know whether or not they need to irrigate the crop, for example. for example, helping the plant in times of greater scarcity or need for water (Farias et al., 2009).

Despite the good performance of Brazilian production, there are still major losses caused mainly by a lack of water during plant development, especially in the germination and grain filling phases. In the 2018/2019 crop year, losses of around 37 per cent of crop yield were estimated in the state of Paraná, according to data from previous harvests (Paraná, 2019). In the 2021/2022 harvest year, several producers recorded losses on their soya crops in the state of Rio Grande do Sul, according to a report produced in partnership between Emater/RS and RTC - Fecoagro/RS and released by the Secretariat of Agriculture, Livestock and Rural Development - SEAPDR, losses of around 26.5% of the state's total production were estimated, which, converted into reais, comes to around 14 billion, due to the severe drought faced during the period. Around 207,000 properties were affected across a wide range of activities (SEAPDR/RS, 2022).

For the seed market, different companies and their respective research laboratories are working to develop cultivars that are more resistant to prolonged drought. Different resistant soya varieties are already circulating in the market, and each year new ones are introduced that are more efficient at dealing with this and other damaging factors. However, it is still very difficult for producers to prevent reasonable losses in production, especially when there is a lack of water during the periods of greatest need, which include the germination and vegetative establishment phases (V-1), flowering (R-1) and initial grain filling (R-5). (Sentelhas et al., 2015).

One alternative that can help when there is little rainfall is supplementary irrigation, more precisely irrigation by reel and sprinkler or centre pivot. However, for the producer, this is a bold investment with a high cost, as both require large reserves of water (reservoirs, dams, etc.), as well as good management of this resource, as misuse or excess can lead to significant losses. Analysing all the factors that involve a lack of water in soya production, one can see that various studies and technologies are being carried out with the main aim of improving production efficiency or resisting these complications, always with the aim of helping the producer and overall production, whether inside or outside Brazil.

* **WHEAT**

Wheat is one of the most important crops globally, with water requirements varying between 450 mm and 650 mm during its growth cycle. Proper water management, considering both scarcity and excess, is crucial to guaranteeing high productivity and grain quality. As well as focusing on water efficiency, the researchers are also studying the colour of the leaves for decorative purposes.

Water availability is vital for wheat production, influencing all stages of the plant's growth cycle. Inadequate management of water resources, whether through scarcity or excess, can seriously jeopardise productivity and grain quality. Each stage of wheat development, from germination to grain filling, requires a specific amount of water.

During the growth cycle, wheat needs approximately 450 mm to 650 mm of water. The grain filling phase is the most sensitive, as a lack of water can drastically reduce wheat yield and quality. Water stress, caused by a shortage of water, can impair photosynthesis, limit nutrient absorption and cause plants to ripen early, resulting in grains of lower weight and inferior quality (Berlato et al., 1986; Bergamaschi et al., 1999).

Excess water also presents challenges. Soil waterlogging can lead to root suffocation, preventing oxygen absorption and increasing susceptibility to disease. This not only jeopardises plant development, but can also cause soil structures to deteriorate, making it difficult for roots to anchor themselves and for nutrients to be absorbed efficiently (Marchezan et al., 2007).In order to manage water efficiently, it is essential to know the total evapotranspiration, which includes water loss from both the soil and the plant.

Evapotranspiration varies according to climatic conditions and the stage of wheat development, and is a fundamental metric for guiding irrigation and other water management practices (Costa, 2001). Knowing these needs helps growers decide when and how much to irrigate, avoiding both water stress and excess moisture.

Research and the development of new agricultural technologies are crucial to optimising the use of water in wheat production. Institutions such as Embrapa have invested in studies to identify the specific water requirements of different wheat cultivars, with the aim of maximising water use efficiency and improving productivity and grain quality. These technological advances include the development of varieties that are more resistant to adverse conditions and the implementation of more precise and efficient irrigation systems.

In conclusion, managing water availability is a determining factor for successful wheat production. Understanding the plant's water needs and applying appropriate water management practices are essential to ensuring a high-quality harvest and minimising losses. The combination of scientific knowledge, technology and innovative agricultural practices allows producers to tackle water-related challenges and ensure the sustainability of wheat production.

REFERENCES

BERGAMASCHI, H.; BERLATO, M. A.; MATZENAUER, R.; FONTANA, D. C.; CUNHA, G. R.; SANTOS, M. L. V.; FARIAS, J. R. B.; BARNI, N. A. Agrometeorology applied to irrigation. 2. ed. Porto Alegre: Ed. UFRGS, 1999. 125 p.

BERGAMASCHI, H.; COSTA, M. M. Effects of Water Stress on Plants.Agronomy, 1999.

BERLATO, M. A.; BERGAMASCHI, H. Water Requirement and Evapotranspiration. Brazilian Journal of Agrometeorology, 1986.

COSTA, A. R. As relações hídricas das plantas vasculares. Portugal: Ed. Universidade de Évora, 2001. 75 p. (Academic text).

COSTA, L. C. Irrigation Management for Wheat. Embrapa Technical Bulletin, 2001.

FARIAS, J. R. B.; NEUMAIER, N.; NEPOMUCENO, A. L. Soya. In: MONTEIRO, J. E. B. Agrometeorologia dos cultivos: o fator meteorológico na produção agrícola. Brasília, DF: INMET, 2009. p. 263-277.

HUSSAIN, H. et al. Enhanced drought resistance of rice plants with a synthetic trehalose gene. Journal of Plant Physiology, v. 171, n. 14, p. 1287-1296, 2014.

MANGRAVITE, H. M.; NOELLE, R. J.; MARUYAMA, W. I.; BAVARESCO, A. Relationship between grain yield and yield components in irrigated rice genotypes. Semina: Ciências Agrárias, v. 42, n. 3, p. 1129-1140, 2021.

MARCHEZAN, E.; MAGALHÃES, A. M. T.; SILVA, P. R. F. Impacts of the Excess of Water in the Soil. Brazilian Agricultural Research, 2007.

PARANÁ. Department of Agriculture and Supply. Department of Rural Economy. Estimates of Estimates. 2019. Available at: <http://www.agricultura.pr.gov.br/arquivos/File/deral/pss.xls>. Accessed on: 20 May 2019.

PETRINI, J.A.; FAGUNDES, P.R.R.; MAGALHÃES Jr, A.M. de; GOMES, A. da S.; ANDRES, A. Strategy for reducing water use in irrigated rice: super-early cultivar BRS Atalanta. Documents, 231. Pelotas: Embrapa Clima Temperado, 2008. 17p.

RIO GRANDE DO SUL. Department of Agriculture, Livestock and Rural Development. Drought report – No. 01/2022. 2022. Available at: <https://www.agricultura.rs.gov.br/upload/arquivos/202204/04103720-relatorio- estiagem-01.pdf>. Accessed on: 02 July 2024.

SENTELHAS, P. C.; BATTISTI, R.; CÂMARA, G. M. S.; FARIAS, J. R. B.; HAMPF, A. C.; NENDEL, C. The soybean yield gap in Brazil - magnitude, causes and possible solutions for sustainable production. Journal of Agricultural Science, v. 1, p. 1-18, 2015.

SINGH, R. K.; SINGH, U. S. Rice breeding for yield improvement under changing climate. Plant Breeding, v. 140, n. 2, p. 215-227, 2021.

SORENG, R. J.; PETERSON, P. M.; ROMASCHENKO, K.; DAVIDSE, G.; TEPE, E.

J.; VALDÉS-REYNA, J. A worldwide phylogenetic classification of the Poaceae (Gramineae) II: An update and a comparison of two 2015 classifications. Journal of Systematics and Evolution, v.55, n.4, p.259-290, 2017.

SOYA PRODUCTION TECHNOLOGIES. Published by: Embrapa soya - Londrina/PR, 2020, Chap. 2 - Eco-physiology of soya, p. 41-43, 2020.

CHAPTER 4

FERTILISING RICE, WHEAT AND SOYA CROPS

Amanda Sanches Thainara Oliveira Viviane Trassante
Joice Fernanda Lübke Bonow Andréa Bicca Noguez Martins

FERTILISER RECOMMENDATIONS BASED ON CHEMICAL SOIL ANALYSIS

1. FERTILISING SOYA BEANS

Soya is one of the world's top grain producing crops. It is grown almost everywhere in Brazil, achieving high levels of productivity. This crop is the main source of income for the country and for rural producers, leading the ranking of the most exported products for over 22 years, that is, since Brazil began to record and publicise foreign sales data. In recent years it has been gaining even more ground due to the almost guaranteed profitability of the crops (POPOV, 2019).

In Brazil, soya occupied an area of 38.2 million hectares in 2020 (EMBRAPA, 2020). The adoption of good agricultural management practices, improved cultivars and modern technology has led Brazilian crop yields to experience a new level of productivity in recent years. It is important to note that efficient soil fertility management is a decisive factor in defining the productivity of soya crops (TANAKA, 2018).

Soya has become one of the most important crops in recent decades, playing a key role in global food security. Soya produces more protein per hectare than any other major crop. In the Cerrado, soya accounts for 90% of the biome's agriculture; to give you an idea, in the 2017/2018 harvest, more than half (52%) of the soya grown in Brazil was concentrated in the Cerrado (VENCATO, 2020).

1.1 The importance of soya for agribusiness

Over the years, soya production has gained great economic value in agribusiness, where it is growing more and more. In Brazil, soya is the main grain for commercialisation, and its growth is due to the consolidation of the oilseed as an important source of vegetable protein, especially to meet the great demand from sectors focused on animal products; the supply of new technologies that help not only to expand but also to exploit soya production in other parts of the country, with a view to new productions that help not only for those who grow it, but also in the economy and in the supply of the product, whether fresh or derived to meet population demand (HIRAKURI and LAZZAROTTO, 2014).

In view of the above, the soya market is largely geared towards various food sectors, from marketing in natura, bran, oils, by-products and even biofuels, where it fits into the industrial biodiesel sector. So, from any angle of agribusiness, we can see the great advances that have taken place over the years, and when it comes to soya, its expansion is largely due to the increased importance of grains and their derivatives for the domestic and foreign markets. It is worth noting that with the advance of technologies and good production management that provide sustainability for Brazil, better governance has helped to maximise producer profits

(GAZZONI, 2012).The expansion of soya in Brazil took place in the mid-1970s with an interest in the oil industry. In 1975, the crop was produced using cultivars and techniques that came from outside the country, specifically the United States, but large-scale cultivation only worked in the southern regions, where the cultivars had environments with conditions similar to their real country of origin. As a result, the tropical cultivar was created for the tropical regions of Brazilian soil and, soon after, new cultivars were created that could be adapted to other locations, bringing stability. It is worth mentioning that cultivation has brought the country an increase in the seed market, providing stability for greater economic exploitation in regions where the land had nothing but forests and savannahs (PONTES et al., 2009).

In this sense, it can be considered that the soya production chain has helped and still helps the Brazilian economic sector, where there has been exploration for the implementation of cultivation in other regions with cultivars created for better adaptation and better production, emphasising that, in addition to helping the country's economic sector, it also helps the regional sector of the chosen location and the producer responsible for cultivation using the appropriate cultivars and with more effective management techniques for planting (HIRAKURI and LAZZAROTTO, 2014).

1.2 Soya Crop (Glycine Max)

Soya is a plant of Asian origin and its cultivation is completely different compared to five millennia ago, when it was a creeping plant grown near rivers and lakes, known as wild soya (MOZZAQUATRO et al., 2017). Over the years, its evolution began with the emergence of new plants that originated from the natural mating of two wild soya beans, which were also domesticated and improved by the Chinese (MORAES et al., 2021).

As mentioned, it is a crop that originated on the Asian continent, more specifically in China, is very rich in protein and was introduced into agriculture over 5,000 years ago. The first record of soya beans was in the book "Pen Ts'ao Kong Mu", which contained descriptions of the plants in China for Emperor Sheng-Nung. It was not introduced to the West until around the 15th century, on the European continent with a totally different purpose to China, where instead of being used for food, it was used for decoration in the botanical gardens of France, England and Germany (BERTRAND et al., 1987).

They go on to say that for the Chinese at that time, soya was one of the mainstays of agriculture, along with rice, wheat, barley and millet. Its role in society was very important in the country, as it was used as an object for borrowers and was also one of the main foods accumulated by Buddhist monks. The crop is typical of temperate countries, has been tropicalised and is currently one of the most established crops in the country. Its cultivation began in the southern states in the mid-1970s, progressing to an expansion in the cerrado region from the 80s onwards. By 1990, the areas where soya was grown had already made great progress in the central part of the country and was closely associated with the expansion of soya plantations in the cerrado. As the years went by and cultivation progressed, Brazil became a major world exporter in 2003 and 2004, accounting for 8% of exports respectively (DOMINGUES et al., 2014).

1.3 Soya cultivars

The increase in cultivated areas and productivity is due to genetic improvement. Its development, for the production of new cultivars, tends to promote improvements in the production chain with regard to the increase and stability of the crop. However, it is worth emphasising the importance of evaluating these cultivars in the producing regions, since the genotypes introduced can have a positive effect on the plant's development in a given location or be unviable in other locations (CORREIA et al., 2017). According to Silva and Duarte (2006), there are evaluations aimed at identifying cultivars that have greater stability in the field. development and predictable responses to environmental variations, which will depend on the growing environment that can have a positive or negative influence on the plant's agronomic characteristics. Cultivars, whether they are determinate, semi-determinate or indeterminate, have good production potential. Indeterminate cultivars tend to have a longer reproductive process in which they tend to recover better from the effects of water stress, either due to water shortage or excess. They require greater care when it comes to defoliation and pest control during this period (THOMAS, 2018). In order to choose the best cultivar that will show the best development and performance in each region, a series of tests should be carried out with other cultivars, making comparisons based on production characteristics (CORREIA et al., 2017).

1.4 EARLY FERTILISATION

Fertiliser management is a set of planned and organised practices with the aim of efficiently and economically supplying fertilisers to crops. Correct fertiliser management consists of a set of decisions on how, from what source, when and in what quantity to use fertilisers in order to achieve the greatest technical and economic efficiency, which, together with pest and disease control, contributes to increased productivity (FERRARI, 2019). A common factor among most of the world's record-breaking farmers is their ability to identify a highly productive soil environment. Structured soil, high levels of organic matter, low penetration resistance, high water infiltration and storage capacity, combined with high levels of nutrients characterise environments with high productive potential (HOEFT, 2020). Conservation practices such as the adoption of the no-till system, together with crop rotation, bring about increases in productivity over time, improving water storage in the soil. These steps are important for obtaining high yields, where the management of the previous crop will have a major influence on subsequent crops (ERNANI, 2020). Early fertilisation of soya crops has become an increasingly discussed alternative that raises questions. In this form of fertilisation, the recommended dose is applied in advance, partially or in full, to the main summer crop, while it is still on the preceding crop, and it can be applied by spreading or incorporation (CAMARA, 2020). This method is used to reduce the time it takes to refuel the seeder-fertiliser, reduce the number of tractor-seeder combinations and consequently reduce production costs (ANGHINONI, 2018). Another positive aspect of early fertilisation compared to fertilisation at sowing is the increase in the formation of straw, increasing organic matter in the production system, improving soil conditions, helping to preserve it, maintaining more moisture and enabling the recycling of nutrients through the mineralisation of organic matter, which makes them available to the next crop (ROSOLEM, 2021). According to Vitti

(2018), higher levels of fertiliser on soya crops provide greater accumulation of dry matter and greater release of potassium for the subsequent crop. The straw is an important nutrient reservoir for subsequent crops, making large quantities of nutrients available in the short term. Soya crops, in soils with adequate fertility levels, do not respond to increases in fertiliser levels, and fertiliser can be applied early. In soybean production systems where there is the possibility of a second summer crop, they would benefit from faster planting operations in both crops, favouring better use and distribution of water in the crop cycle before the end of the rainy season (MENDES, 2018). According to Cantarella (2020), in order to increase the productivity of a crop that is already so highly exploited, it is necessary to take into account various factors linked to its development that could increase or reduce its productivity. Soya is a plant that is sensitive to the photoperiod of the environment; changes in this environment cause the vegetative phase to change to the reproductive phase. Delays in sowing cause a reduction in vegetation time and, consequently, in productivity, because by delaying the establishment of the crop, the end of the juvenile period is brought closer to the day when the plant is ready for the reproductive phase, in the adaptation region (MIRANDA, 2021). In addition to the effects on productivity in the summer crop, bringing forward fertilisation to the winter period provides better climatic conditions for the use of nutrients, especially nitrogen, due to lower losses to the atmosphere. In this sense, spreading and anticipating fertiliser brings greater agility to the process, also helping to minimise saline effects in the sowing furrow caused by high doses of fertiliser, especially when using NPK formulations with potassium chloride in their composition (ZANGH, 2018). In addition to better optimisation of the machinery, there is also better absorption of nutrients by the winter crop, generally due to the greater number of plants per hectare, lower fertilisation costs due to the lower value of inputs in this period, increased biomass production, a reduction in the C/N ratio in poaceous plants, and less loss of nitrogen to the atmosphere. The consequences of all this are operational, economic and environmental gains (FANCELLI, 2018). The practice of early fertilisation is common in extensive cultivation systems, where it is necessary to maximise the operational yield when sowing the summer crop in order to make better use of the sowing windows. It is also possible to split the volume of fertiliser added, adding part of the fertiliser to the preceding crop and the rest pre-sowing (CALONEGO, 2018). As well as improving operational performance during sowing, another very important benefit of system fertilisation is the increased uniformity of fertiliser distribution on the crop, especially when system fertilisation is applied to autumn/winter crops such as winter cereals. Because there is less spacing between rows compared to summer crops, fertilising winter crops allows for better distribution of fertilisers in the soil (RAIJ, 2019). In the no-till system, using the traditional fertilisation method, the amount of N recommended at sowing may not be enough to meet the nutritional requirements of the plants in the early stages due to the lack of N in the initial phase caused by the immobilisation effect of mineral N. This means that fertilising with N in the early stages is not enough to meet the nutritional requirements of the plants. Therefore, early fertilisation can be a solution to this problem. N applied early can be momentarily immobilised by organic matter, especially by residues with a high carbon/nitrogen ratio (C/N), but as it was applied early, there is enough time for it to become available for the subsequent crop (HIRAKURI, 2019). There is a mutual benefit between these two techniques. Just as early fertilisation solves the problem caused by nitrogen immobilisation, SPD contributes to improving soil fertility, enabling phosphorus and potassium to be safely applied early. In addition, SPD makes for a more sustainable production system, reducing the release of carbon into the atmosphere due to the straw covering the soil, and also reducing the emission of fossil fuel gases due to the reduction in mechanised operations (MARTINAZZO, 18). Another positive aspect that early fertilisation can bring to SPD is that it leads to greater mass production of cover crops. When SPD is adopted, one way of anticipating fertilisation is to apply part of the fertiliser while the green cover is still growing, which allows these plants to develop better, producing good mulch (ESTEVES, 2020). In a study carried out by Embrapa Solos in Turvelândia, GO, soya beans were planted in the spring and cotton in the summer. Comparisons were made between the fertilisations, which were carried out in advance while soya was still being grown. and in the cotton planting line. The results of the experiment showed that potassium fertilisation during millet cultivation was beneficial, as it stimulated the production of phytomass. The millet responded with a 45 per cent increase in dry matter production. There was also another advantage of the system, as in order to achieve yields close to those with early planting, the producer would have to opt for 2 instalments, which would make the cost of production higher (HIRAKURI, 2019). It's important to note that in the SPD, the green cover will also make nutrient cycling more efficient, reinforcing the idea that the purpose of fertilisation is to maintain soil fertility at satisfactory levels. Thus, we can say that the adoption of SPD, together with the anticipation of fertilisation, enables a systemic action, and not just a one-off action as in the traditional system (ESTEVES, 2020). Another study carried out in the experimental area of the Federal Centre for Technological Education in Rio Verde, in the southwest of Goiás, aimed to compare the productivity of soya crops fertilised with phosphorus (P) and potassium (K) at sowing and in the field. The results showed that there was no difference between the treatments when it came to the weight of 1000 grains, total number of pods per plant and grains per pod. There was a significant difference only between the early P+K fertilisation, which outperformed the early P and control treatments. There was no significant difference between phosphate and potassium fertilisation at sowing and early sowing, which suggests opting for the least costly form of application (GUARESCHI, 2008).

2. FERTILISING WHEAT CROPS

Wheat is a crop of great importance on the world stage, being the second most widely planted crop behind maize. It is one of the most widely consumed cereals in the world and is used for food, animal feed and in industry, with several options for constituent by-products on the market (CARLETTO, 2013).

When it comes to nutrition, nitrogen (N) and potassium (K) stand out in terms of their extraction needs by various crops, especially cereals, and there is an interaction between them that has been studied for mass and grain production. Nitrogen is an essential component for the formation of essential compounds and plant survival, while potassium is related to various biochemical reactions in plant metabolism (VIANA, 2007). These nutrients are essential for the production of green mass in the aerial part of plants, grain production and root biomass, and there can be a decrease or absence of productivity in the presence or absence of one of these nutrients, due to aspects related to soil fertility, climatic conditions and, most importantly, the imbalance of these nutrients (VIANA, 2007). According to Date (2000); Graham and Vence (2000), nitrogen is currently in second place as the biggest factor limiting production, second only to water deficiency. It is fundamental in the formation of the grain's essential compounds, made up of proteins, carbohydrates and gluten, among others. According to Espíndola et al. (2010), the recommendations are wide-ranging and the dynamics of application and utilisation depend on the specifics of the growing conditions, so these recommendations should be adopted with caution. The interaction between N and K can occur in different places and at different times when considering soil-plant, when a sufficient amount of K is applied to increase production, this will be limited by the low amount of N in the soil, this analysis is important for knowledge and food production, interfering in productivity, and the balance provides maximum efficiency and results in the utilisation of these compounds by plants (VIANA, 2007).Among the biochemical functions of K in plants, one of them is the transport of assimilates from photosynthesis, a function in translocation by the phloem, requiring the energy provided by ATP, which is necessary for production, where this fertilisation with K contributes significantly to the assimilation of N by the grains (MALAVOLTA, 2006).Nitrogen fertilisation in wheat is recommended by experts in the field, and is extremely important for meeting the crop's needs and increasing productivity. Its availability is one of the most important factors in the process of plant growth and development, as it is the nutrient with the greatest impact on the production and quality of cereals (VIANA, 2007). There are several sources of nitrogen, but urea is more concentrated, which provides greater interest, ensuring greater consumption in Brazil. Among all the sources, it has the highest nitrogen content (45 %), contributing to a lower cost and ensuring numerous advantages, fast reactions and compatibility with numerous fertilisers and pesticides (YANO, TAKAHASHI, WATANABE, 2005).This nutrient is also essential during the phenological stages of the crop, initially for starting the crop and forming the aphylls, and this need is met by the fertiliser provided at sowing. Later, nitrogen topdressing is indispensable, as it helps to increase the leaf area and is responsible for vegetation, which is reflected in the production of vegetative buds, tillering, the protein content of the grains and productivity (MALAVOLTA, 2006). Another part of the nitrogen required by the crop is after tillering, which is supplied by means of top dressing, because it helps development, preceding the phase called rubberisation, where the ear is still attached, inside the sheath, waiting for the right moment to be released, thus constituting one of the important functions and management practices of grasses, which contributes to better grain filling and quality and increased productivity (SANGOI et al., 2007).

Figure 1: Example of an N recommendation for wheat, with applications in instalments at the times when the crop needs it most

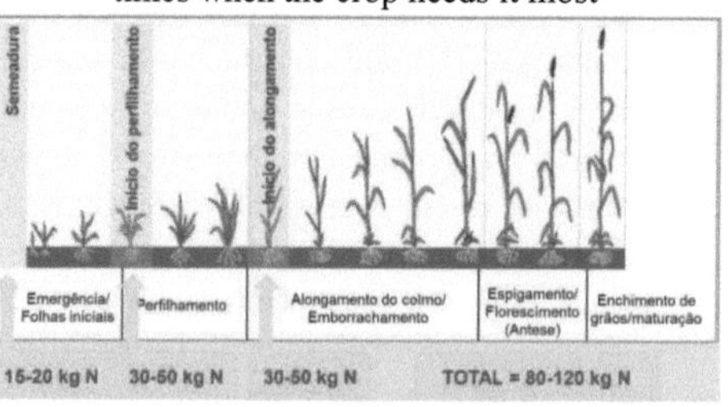

Source: Embrapa Wheat.

In terms of wheat productivity, there is greater success in relation to the quantity of inputs, including nitrogen fertilisation, techniques used during the crop cycle, management and the genetic material of the cultivar planted, as well as expected yield expectations, plant height and soil fertility.However, very low doses of nitrogen limit the production of species that produce tillers, such as rice and wheat. A lack of nitrogen reduces the formation of tillers, which reduces the number of inflorescences and consequently production.

Figure 2: Reduction in tiller number in wheat plants due to N deficiency:

Source: Yara fertilisers

Figure 3: Wheat crop lodging due to excess nitrogen.

Source: UFPR.

As a result of this undesirable situation, in order to reduce losses and get around this problem, the market has come up with new products, more technified nitrogen fertilisers based on protected urea, strategies to increase efficiency and promote lower losses, reducing the import of fertilisers and contamination of the environment and thus maximising productivity, as well as allowing nitrogen fertilisation to be carried out without the need for large amounts of rain or humidity at specific times for its action (PRANDO et al, 2012).

Nitrogen losses are lower in protected urea, which is a product manufactured using principles to protect against losses to the environment. Its structure is protected with a capsule for this purpose, where its less volatile and dispersive form in the soil slowly degrades and gradually becomes available. for the plant, resulting in better utilisation and adequate nutrition, according to the crop's needs, and thus losing less nitrogen to the environment, which occurs in the form of ammonia gas (BONO et al., 2008).

The urea is retained by the gelatinised starch, which guarantees the product great stability and slow degradation of its components inside the soil, promoted by the physical impediment of the polymer, reducing waste and making it available at the most appropriate time during the plant cycle (BONO et al., 2008).

3. TYPES OF FERTILISER FOR RICE CULTIVATION.

Figure 4: Rice

3.1 Nitrogen fertilisation

Nitrogen (N) promotes plant growth by increasing the number of tillers, the number of particles and the number of grains per particle. Phosphorus (P) is necessary for tillering, grain formation (filling) and grain quality (BARBOSA FILHO, 1989; CASTRO et al., 2013). Potassium (K) is one of the essential nutrients for the growth of rice, acting on the plant's photosynthesis, contributing to the opening and the closing of leaf stomata, as well as being responsible for the transport of soluble carbohydrates within the plant (SANTOS, 2018). However, in addition to all these effects, K has the ability to strengthen the cell walls of the stalk with lignin, conferring greater resistance to lodging, diseases and pests in rice (BARBOSA FILHO, 1989). Nitrogen is the second most extracted nutrient and the most exported by the rice crop (FORNASIERI FILHO; FORNASIERI, 2006), and is a constituent of numerous organic compounds such as amino acids, nucleic acids and protein (EPSTEIN, 1975). Nitrogen deficiency in rice is relatively common in Brazil, which is related to factors such as: low organic matter content in the soil; losses through leaching and volatilisation; reduced use of nitrogen fertilisers; nutritional imbalance; water deficiency (EPSTEIN, 1975; RAIJ, 1991; FAGERIA; SANT'ANA; MORAIS, 1995). It is one of the main factors involved in productivity (FAGERIA; BALIGAR, 2001; FAGERIA; BARBOSA FILHO, 2001) and in improving the nutritional quality of rice grain (FERRAZ JUNIOR et al., 1997). When supplied in quantities greater than the needs of vegetative growth, close to anthesis, it makes it possible to increase the protein content (CHING; RYND, 1978), with an increase in biological value by positively interfering in the protein fraction of glutelin (CHING; RYND, 1978; FERRAZ JUNIOR et al., 1997). Rice is a staple food for the human population and a source of energy for more than half of the world's population. The cereal is grown on around 160 million hectares, corresponding to approximately 11 per cent of the world's cultivated land (Grisp, 2013). In Brazil, the largest rice producer outside the Asian continent, a large part of the cereal's production comes from irrigated crops in Rio Grande do Sul (RS). With the highest national average productivity of 7,925 kg ha-1 (Embrapa Arroz e Feijão, 2018), irrigated rice plantations in Rio Grande do Sul currently account for 55 per cent of the country's production. area, contributing more than 70 per cent of national cereal production (Conab, 2018).Specifically for rice, nitrogen is the most important nutrient for productivity, providing the greatest responses to fertilisation. However, the agronomic efficiency of the nutrient, i.e. its capacity to increase productivity per unit of nutrient added to the soil, is highly variable due to the complex interaction of factors that determine its utilisation by the crop. The main factors involved are: characteristics of the cultivar; climatic conditions; supply of N and other nutrients in the soil; crop sequence; sowing time and density; irrigation management; weed control; the crop's phytosanitary status and nitrogen fertiliser management (Scivittaro; Machado, 2004).

The input of N via fertiliser is responsible for the increase in grain yield and rice yield components, such as the number of tillers and panicles per unit area (Singh; Pillai, 1996), the number of spikelets per panicle and grain weight (Marzari, 2005). Although rice absorbs nitrogen throughout its cycle, the demands are greater in the tillering and reproductive phases. However, it is in the latter, which begins with panicle initiation, that the plant is most efficient at absorbing N for grain production, since the root system is more developed and consequently has greater potential for absorbing nutrients (Machado, 1993). Therefore, the time and form of nitrogen top dressing applied to rice are strongly associated with the efficiency of the crop's use of the fertiliser nutrient and its response to fertilisation (Scivittaro; Machado, 2004).

3.2 Potash fertilisation

Another important nutrient for rice is potassium (K). It is an essential nutrient, acting in various physiological and biochemical processes in the plant, including osmotic regulation, enzyme activation, pH regulation and cellular ion balance, regulation of stomatal transpiration and transport of photosynthesis products (Dobermann; Fairhurst, 2000). The demand for potassium by irrigated rice is as high as or higher than that for N, and can reach 15.9 kg of K per tonne of grain produced (Buresh et al., 2010). However, the content of this nutrient in the grains is relatively low, indicating that only a small fraction of the nutrient absorbed is exported to the grain (Scivittaro; Machado, 2004), the rest being kept in the system (Anghinoni et al., 2013).Research results show that potassium fertilisation brings various benefits to rice cultivation, including promoting the development of the root system, increasing plant vigour, reducing lodging, increasing resistance to pests and diseases, increasing tillering and dry matter production of the aerial part and roots, increasing the number of full grains and grain weight, among others (Harari et al., 2018). However, the response of irrigated rice to the application of potassium is not as marked as that obtained for nitrogen, especially in flooded cultivation, in which the availability of the nutrient is greater, due to increased diffusion, the displacement of the nutrient from the exchange sites to the soil solution, by the cations $NH4^+$, $Fe2^+$ and $Mn2^+$ (Machado, 1985), and the release of potassium from the non-tradable and structural fractions (Castilhos; Meurer1999a, 1999b; Castilhos et al., 1999).

REFERENCES

BERTRAND, Jean Pierre; LAURENT, Catherine; LECLERCQ, Vincent. The World of Soya. Ed. HUCITEC- Editora da Universidade de São Paulo. São Paulo, 1987.

BONA, D, F; WIETHOLTER, S. Management of the wheat production system for efficient use of fertilisers. Embrapa Trigo, Foz do Iguaçu PR, Feb. 2014.

CARLETTO, R. Agronomic and forage characteristics of dual-purpose wheat subjected to cutting systems in Cv. BRS Umbu. Midwestern State University, Postgraduate Programme in Agronomy, UNICENTRO, Plant Production, Guarapuava, 2013.

CASTILHOS, R. M. V.; MEURER, E. J. Potassium supply for flooded rice in RS soils. In: BRAZILIAN CONGRESS ON IRRIGATED RICE, 1; REUNION OF IRRIGATED RICE CULTURE, 22, 1999, Pelotas. Proceedings...
Pelotas: Embrapa Clima Temperado, 1999b. p. 334-337.
CAVALCANTE, A, J; PRIMIERI, C; RIBEIRO, T, E; DELUCA, R; SILVA, G, W.

Revista cultivando saber, special issue, p-1-14, 2016
DOBERMANN, A.; FAIRHURST, T. A. H. Rice: nutrients disorders and nutrient management. Los Baños: IRRI, 2000. 191 p.

DOMINGUES, M. S. D., BERMANN, C., & SIDNEIDE MANFREDINI, S. A produção

of soya in Brazil and its relationship with deforestation in the Amazon. Revista Presença Geográfica, v.1, n.1, 2014.

EPSTEIN, E. A. Nitrogen acquisition. In: MALAVOTA, R. (Trad.). Mineral nutrition of plants: principles and perspectives. Rio de Janeiro: Livros Técnicos e Científicos; São Paulo: EDUSP, 1975, p. 213-234.

ESPINDULA, M. C.; ROCHA, V. S.; SOUZA, M. A.; GROSSI, J. A. S.; SOUZA, L. T.

Doses and forms of nitrogen application in the development and production of the Wheat crop. Ciências Agrotec, Lavras, v. 34, n. 6, p. 1404-1411, nov-dez, 2010.

FAGERIA, N. K.; BALIGAR, V. C. Lowland rice response to nitrogen fertilisation. Communication Soil Science Plant Analysis, Philadelphia, v. 32, n. 9-10, p. 1405- 1428, 2001.

FAGERIA, N. K.; BARBOSA FILHO, M. P. Nitrogen use efficiency in lowland rice genotypes. Communication Soil Science Plant Analysis, Philadelphia, v. 32, n. 13-14, p. 2079-2089, 2001.
FAGERIA, N. K.; SANT`ANA, E. P.; MORAIS, O. P. Response of favoured rainfed rice genotypes to soil fertility. Pesquisa Agropecuária Brasileira, Brasília, v. 30, n. 9, p. 1155-1161, 1995.

FERRAZ JUNIOR, A. S. L.; SOUZA, S. R.; FERNANDES, M. S.; ROSSIELLO, R. O.

P. Nitrogen use efficiency for grain and protein production by rice cultivars. Pesquisa Agropecuária Brasileira, Brasília, v. 32, n. 4, p. 435-442, 1997.

HARTARI, S.; SURYONO; PURNOMO, D. Effectiveness and eficiency of potassium fertilizer applicatîon to increase the production and quality of rice entisols. Earth and Environmental Science, v. 142, 012131, 2018. doi:10.1088/1755-1315/142/1/01231

HIRAKURI, M. H., & LAZZAROTTO, J. J. (2014). Soya agribusiness in the global and Brazilian contexts. Embrapa Soja-Documentos (INFOTECA-E), 2014.

https://bdm.unb.br/bitstream/10483/4129/1/2012_FabioCaribedeAraujoGalvao.pdf
https://repositorio.animaeducacao.com.br/bitstreams/d44b0401-1a92-44ff-989e-c0974088f1be/download

https://repositorio.ifgoiano.edu.br/bitstream/prefix/3825/7/tcc_Jo%C3%A3o%20Pedro%20Cruvinel%20Guimar%C3%A3es.pdf
MACHADO, M. O. Fertilisation and liming for irrigated rice in Rio Grande do Sul. Pelotas: EMBRAPA-CPATB, 1993. 63 p. (EMBRAPA-CPATB. Research Bulletin, 2)

MACHADO, M. O. Soil characterisation and fertilisation. In: EMBRAPA-CPATB. Fundamentals for growing irrigated rice. Campinas: Cargill Foundation, 1985. p. 129-179

MACHADO, M. O.; ZONTA, E. P.; FAGUNDES, P. R. R.; TERRES, A. L. S. Doses and

nitrogen application times in four irrigated rice genotypes, in three successive harvests. 2000. Unpublished data

MALAVOLTA, E. Manual de nutrição mineral de plantas. São Paulo: Editora Agronômica Ceres, 2006. 638 p.

MORAES, G. N., LEMANSKI, M. C., JULIEN, M. Y. C., CRUVINEL, M. E. M., & DE REZENDE, S. P. (2021, August). SOJA: THE CROP THAT MOVES BRAZIL. In Proceedings of the State Colloquium on Multidisciplinary Research (ISSN-2527-2500) & National Congress on Multidisciplinary Research, 2021.

MOZZAQUATRO, E. M. S. S., ALMIRAO, D. D. O., RIGHI, A. P., & LOPES, J. C. D.

S. Economic viability of soya cultivation on a rural property. REVISTA CONGREGA-MOSTRA DE TRABALHOS DE CONCLUSÃO DE CURSO-ISSN 2595- 3605, v.1, p. 806-824, 2017.

OLIVEIRA, C. Nitrogen management in wheat cultivation and symptoms of nitrogen deficiency and excess in plants. Feb-2023.

POPOV, D. Soya: everything you need to know about production in Brazil. 2019.

Available at: Accessed on: 29 Apr. 2023.
PRANDO, A. M.; ZUCARELI, C.; FRONZA, V.; OLIVEIRA, E. A. P.; PANOFF, B.

Forms of urea and doses of nitrogen in top dressing on the physiological quality o f wheat seeds. Revista Brasileira de Sementes, v. 34, n. 2, p. 272-279, 2012.

SANGOI, L.; BERNS, A. C.; ALMEIDA, M. L.; ZANIN, C. G.; SCHWEITZER, C. Agronomic characteristics of wheat cultivars in response to the timing of fertilisation nitrogen top dressing. Ciência Rural, Santa Maria, v. 37, n. 6, p. 1564-1570, nov- dez. 2007.

TANAKA, R. T.; MASCARENHAS, H. A. A.; BORKERT, C. M. Nutrição mineral da soya. Piracicaba: Potafos, 2018. p. 105-135

VENCATO, A. Z. Brazilian Soya Yearbook 2020. Santa Cruz do Sul: Gazeta Santa Cruz, p. 144, 2020.

VIANA, E. M. Interaction of nitrogen and potassium on nutrition, chlorophyll content and nitrate reductase activity in wheat plants. University of São Paulo School of Agriculture "Luiz de Queiroz", ESALQ, Soils and Plant Nutrition, Piracicaba, 2007.

YANO, G. T.; TAKAHASHI, H. T.; WATANABE, T. S. Evaluation of nitrogen sources and top dressing times for wheat cultivation. Semina: Ciências Agrárias, Londrina, v. 26, n. 2, p. 141-148, 2005.

ZAGONEL, J.; FERNANDES, E. C. Doses and times of application of growth reducer affecting Wheat cultivars in two doses of Nitrogen. Planta Daninha, Viçosa, v. 25, n. 2, p. 331-339, 2007.

CHAPTER 5

MAIN DISEASES IN SOYA, WHEAT AND RICE CROPS

Daniela Cunha Thuany Pereira
Joice Fernanda Lübke Bonow Andréa Bicca Noguez Martins

- **SOYA (GLYCINE MAX L.)**

Figure 1: Soya beans

Source: DNAgro - https://dnagro.com.br/cultura/soja/ (Accessed on: 22 July 2024).

Currently, the crop with the largest cultivated area is soya. This can be explained by issues inherent to the species, such as high energy efficiency and hardiness (FERREIRA, 2014) and, consequently, great worldwide commercial interest. The beans produced are widely used for animal feed and vegetable oil extraction, meaning that soya beans are an important product for other production chains, as well as being consumed directly for human consumption. Brazil is currently the world's largest soybean producer, with production exceeding 135 million tonnes (CONAB, 2022). However, there are several diseases that affect this crop in Brazil, and in order to have good control we must first identify these diseases (CUNHA, D.P).

1. SOME DISEASES CAUSED BY FUNGI:

1.1 ANTHRACNOSE (COLLETOTRICHUM SPP.).

Figure 2: Anthracnose on the pods.

Source:

https://blogs.canalrural.com.br/embrapasoja

/2020/09/15/anthracnose-another-soybean-disease/ (Accessed on: 22 July 2024).

Figure 3: Anthracnose on the plant.

Source:

https://blogs.canalrural.com.br/embrapasoja

/2020/09/15/anthracnose-another-soybean-disease/ (Accessed on: 22 July 2024).

1.1.1 SYMPTOMS:

It can cause seedling death and black spots on the veins of leaves, stems and pods. There may be total pod drop or seed deterioration when harvesting is delayed. Infected pods at stages R3-R4 take on a dark brown to black colour and become twisted; on pods in grain, the lesions begin as anasarca streaks and develop into black spots. The infected parts usually have several black dots, which are the fungus's fruitings (acervuli) (EMBRAPA, 2023).

1.1.2 DEVELOPMENT CONDITIONS:

Anthracnose is a disease that affects the initial stage of pod formation and occurs more frequently in the Cerrado region, due to the high rainfall and high temperatures. In rainy years, it can cause total loss of production, but more often it causes a reduction in the number of pods, inducing the plant to retain its leaves and turn green. The use of infected seeds and nutritional deficiencies, especially potassium, also contribute to the disease's greater occurrence.

Seeds from crops that have been delayed in harvesting due to rain may have higher infection rates (EMBRAPA, 2023).

1.1.3 CONTROL:

It is recommended to use healthy seed, seed treatment, crop rotation, row spacing and stands that allow good aeration of the crop and proper soil management, especially with regard to potassium fertilisation (EMBRAPA, 2023).

1.2   ASIAN RUST (PHAKOPSORA PACHYRHIZI).

Figure 4: Soya leaf with rust

1.2.1 SYMPTOMS:

They can appear at any stage of plant development. The first symptoms are characterised by tiny dots (no more than 1 mm in diameter) darker than the healthy leaf tissue, greenish to greenish-grey in colour, with a corresponding protuberance (uredia) on the lower page of the leaf. The uredia take on a light to dark brown colour and open into a tiny pore, expelling hyaline spores that accumulate around the pores and are carried by the wind (EMBRAPA, 2023).

1.2.2 DEVELOPMENT CONDITIONS:

The infection process depends on the availability of free water on the leaf surface, requiring at least 6 hours, with maximum infection occurring between 10 and 12 hours of leaf wetness. Temperatures between 18°C and 26.5°C are favourable for infection. The sooner defoliation occurs, the smaller the grains will be, consequently, the greater the loss of yield and quality (green grain). American rust (P. meibomiae) is recognised as having little impact on yield; P. pachyrhizi is more aggressive and can cause significant losses (EMBRAPA, 2023).

1.2.3 CONTROL:

Chemical control with fungicides formulated as a mixture of different chemical groups has proved effective. The fungicide should be applied preventatively or at the first symptoms of the disease. Sowing should be done at the beginning of the indicated season, preferably using early cultivars and complying with the sanitary void [eliminating volunteer soya plants (guaxa or tiguera) in the off-season], to reduce inoculum in the following season; avoid sowing in the off-season. Resistant cultivars are available for some regions of Brazil; however, they do not dispense with the use of fungicides, since virulent populations can be selected due to the variability of the pathogen (EMBRAPA, 2023).

1.3 BROWN SPOT (SEPTORIA GLYCINES)

Figure 5: Soya leaf with brown spot:

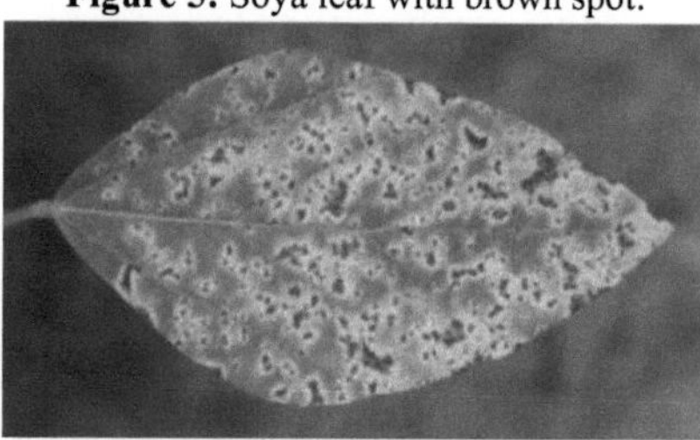

1.3.1 SYMPTOMS:

The first symptoms appear around two weeks after emergence, as small, reddish-brown, angular dots or spots on the unifoliolate leaves. In favourable situations, the disease can reach the first trifolia and cause severe defoliation. On the leaves, brown spots appear, smaller than 1 mm in diameter, which evolve to form spots with yellowish halos and an angular centre, brown in colour on both sides, measuring up to 4 mm in diameter. diameter. In severe infections, it causes defoliation and early ripening (EMBRAPA, 2023).

1.3.2 DEVELOPMENT CONDITIONS:

The fungus survives on crop remains. Infection and the development of the disease are favoured by warm and humid conditions. The spores are dispersed by water and wind. The fungus needs a minimum wetting period of 6 hours and temperatures between 15°C and 30°C to develop symptoms, with an optimum of 25°C. (EMBRAPA, 2023).

1.3.3 CONTROL:

Because the fungus survives in crop remains, the most efficient control is achieved by crop rotation, accompanied by improving the physical and chemical conditions of the soil, with an emphasis on potassium fertilisation. In rainy years, control can be achieved by applying fungicide (EMBRAPA, 2023).

1.4   WHITE MOULD (SCLEROTINIA SCLEROTIORUM)

Figure 6: Soya plant with white mould:

1.4.1 SYMPTOMS:

Watery spots that evolve to a light brown colour and then develop an abundant formation of dense, white mycelium. The fungus is capable of infecting any part of the plant, but infections often start from the fallen petals in the leaf axils and lateral branches. Occasionally, wilting and drying symptoms can be observed on the leaves. Within a few days, the mycelium turns into a black, rigid mass, the sclerotium, which is the fungus's form of resistance. Sclerotia vary in size, and can be formed both on the surface and inside the infected stem and pods (EMBRAPA, 2023).

1.4.2   DEVELOPMENT CONDITIONS:

The most vulnerable stage of the plant is from full bloom to the start of pod formation. High relative humidity and mild temperatures favour the development of the disease. Sclerotia that fall to the ground, under high humidity and temperatures between 10°C and 21°C, germinate and develop apothecia on the soil surface. These produce ascospores which are released into the air and are responsible for infecting the plants. Transmission by seed can occur either through dormant (internal) mycelium or sclerotia mixed with the seeds. Once introduced into the area, the pathogen is difficult to eradicate (EMBRAPA, 2023).

1.4.3   CONTROL:

Avoid introducing the fungus into the area by using certified seed free of the pathogen. Treat the seed with a mixture of contact fungicides and benzimidazoles. In areas where the disease occurs, it is necessary to combine several strategies: direct sowing on grass straw, rotating/succession soya with resistant species such as maize, sorghum, millet, white oats or wheat; eliminating the fungus' host plants; fertilising appropriately; increasing the spacing between rows, reducing the population to the minimum recommended. Fungicide applications can be made at the start of flowering and during flowering (EMBRAPA, 2023).

1.5 TARGET SPOT AND CORYNESPORA ROOT ROT (CORYNESPORA CASSIICOLA)

Figure 7: Soya plant contaminated with target spot

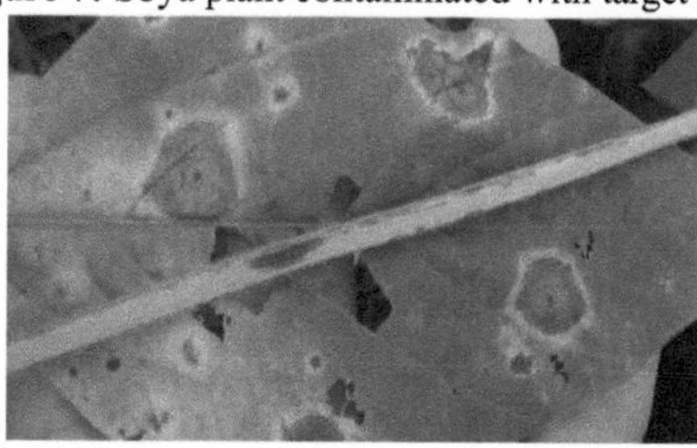

1.5.1 SYMPTOMS:

The lesions begin as brown spots with a yellowish halo, developing into large circular spots, light to dark brown in colour, up to 2 cm in diameter. The spots usually have a dark dot in the centre, similar to a target. Susceptible cultivars can suffer severe defoliation, with reddish-brown spots on the stems and pods. The fungus also infects roots (EMBRAPA, 2023).

1.5.2 DEVELOPMENT CONDITIONS:

The fungus is found in practically all soya growing regions in Brazil. It is apparently native and infects a large number of native and cultivated plants. It can survive on crop remains and infected seed. High relative humidity is favourable for leaf infection (EMBRAPA, 2023).

1.5.3 CONTROL:

The use of resistant cultivars, seed treatment, crop rotation/succession with maize and other grass species and fungicide control are recommended (EMBRAPA, 2023).

1.1 POWDERY MILDEW (ERYSIPHE DIFFUSA)

Figure 8: Soya plant contaminated with powdery mildew

1.6.1 SYMPTOMS:

Erysiphe diffusa is an obligate parasite that develops throughout the aerial part of the plant. It has a thin whitish coating made up of mycelium and powdery spores. On the leaves, over time, the white colour of the fungus changes to greyish-brown and, under conditions of severe infection, it can cause drought and premature leaf fall (EMBRAPA, 2023).

1.6.2 DEVELOPMENT CONDITIONS:

Infection can occur at any stage of the plant's development, but is most common at the beginning of flowering. Conditions of low relative humidity and mild temperatures (18°C to 24°C) are favourable for the fungus to develop (EMBRAPA, 2023).

1.6.3 CONTROL:

The most efficient method of control is the use of resistant cultivars. Chemical control can be used through the application of fungicides (EMBRAPA, 2023).

2 DISEASES CAUSED BY BACTERIA:

2.1 BACTERIAL BLIGHT (PSEUDOMONAS SAVASTANOI PV. GLYCINEA)

2.1.1 SYMPTOMS:

It is common on the leaf, but can attack the stem, petiole and pod. It starts with watery spots, semi-transparent when observed against the light, which necrotise and agglutinate, forming large areas of dead tissue. There may or may not be a wide yellowish halo around the spots at mild temperatures, and narrow or absent at temperatures above 27 °C. Observation of the black spots with irregular edges on the lower page of the leaf allows the disease to be diagnosed in the humid hours of the morning, due to the presence of a shiny film, which is the exudate of the bacteria. Severe attacks cause tearing of the leaf's internerval spaces and leaf fall.

2.1.2 DEVELOPMENT CONDITIONS:

Infected seed and the remains of previous soya crops are the initial sources of inoculum. Infected seed shows no symptoms. The disease is favoured by high humidity, especially rain with wind, and at mild temperatures (20 °C to 26 °C). On dry days, thin scales of the bacterium's exudate are spread on the crop, but for infection to take place there needs to be a film of water on the leaf surface. The bacterium penetrates the leaf through stomata or wounds.

2.1.3 CONTROL:

There are no recommended control measures for this disease.

3 DISEASES CAUSED BY VIRUSES:

3.1   SOYBEAN MOSAIC VIRUS (SMV)

Figure 9: Soya beans with mosaic disease:

Source:

3.1.1   SYMPTOMS:

Infected plants have stunted trifoliate leaves with some blisters and a mosaic distributed irregularly over the leaf limb. Maturation is delayed and it is common to find green plants in the middle of already matured plants. Susceptible genotypes produce seeds with spots (coffee spot). These spots are brown or black, depending on the colour of the hilum. There are, however, susceptible genotypes that do not produce spotted seeds. Seeds without spots can transmit the virus and produce infected seedlings. However, not all stained seeds give rise to infected seedlings (EMBRAPA, 2023).

3.1.2   DEVELOPMENT CONDITIONS:

The common soya mosaic virus was introduced into Brazil via infected seeds and is distributed in all regions where soya is grown. It is transmitted by aphids from host plants. Climatic conditions that favour aphid populations contribute to a higher incidence of the virus in the field (EMBRAPA, 2023).

3.1.3   CONTROL:

As with other plant viruses, the most efficient way to control this disease is through resistant cultivars (EMBRAPA, 2023).

3.2 STEM NECROSIS (COWPEA MILD MOTTLE VIRUS - CPMMV)

Figure 10: Soya beans with stem necrosis:

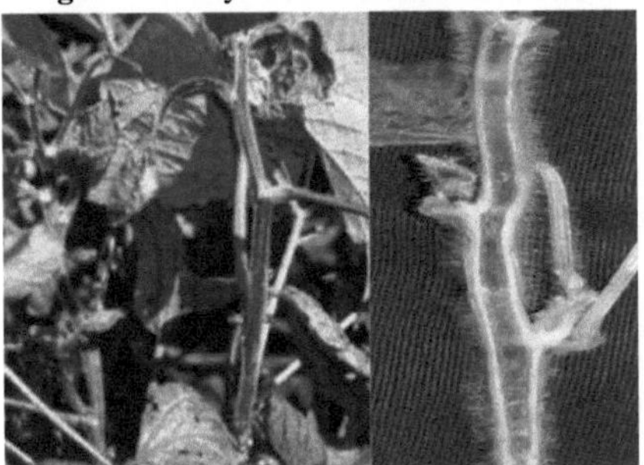

Source:

3.2.1 SYMPTOMS:

During flowering and at the start of pod formation, the symptoms become evident with the appearance of bud burn and stem necrosis, when the plants eventually die. Stems cut lengthways show blackening of the pith. Plants that don't die show severe dwarfism and deformed leaves. Infected plants can produce deformed pods and small grains (EMBRAPA, 2023).

3.2.2 DEVELOPMENT CONDITIONS:

The virus is transmitted by the whitefly (Bemisia tabaci). Any condition that favours the development of the whitefly population also favours the appearance of the disease, as long as there is an infected host plant available. Desmodium tortuosum and Arachys pintoi are host plants for this virus in Brazil (EMBRAPA, 2023).

3.2.3 CONTROL:

Use resistant cultivars. Whitefly control is not effective. In addition to the difficulties of controlling this insect, non-persistent transmission favours the spread of the virus in soya fields (EMBRAPA, 2023).

4. DISEASES CAUSED BY NEMATODES:

4.1 CYST NEMATODE (HETERODERA GLYCINES)

Figure 11: Soya roots with cyst nematodes:

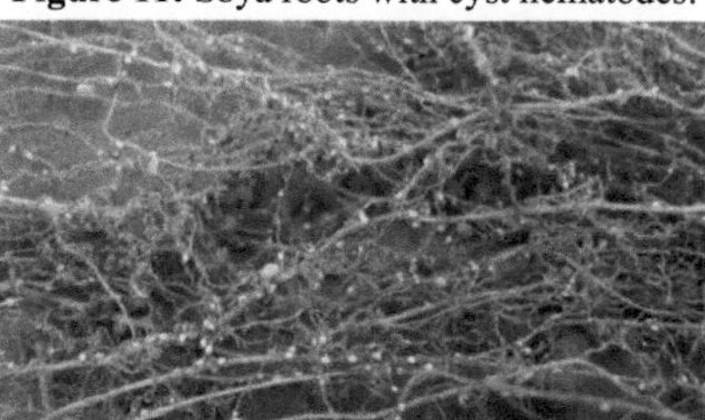

Source:

4.1.1 SYMPTOMS:

The nematode penetrates the plant's roots and hinders the absorption of water and nutrients, causing a reduction in the size and number of pods, chlorosis and low productivity. Symptoms appear in clusters and, in many cases, the plants die. The root system is reduced and there are tiny, slightly elongated lemon-shaped females, white to yellowish in colour. When the female dies, her body turns into a resistant, dark brown structure full of eggs, called a cyst, which detaches from the root and remains in the soil. Diagnosis requires analysing soil and/or root samples in a nematology laboratory (EMBRAPA, 2023).

4.1.2 DEVELOPMENT CONDITIONS:

The cyst can remain in the soil for more than eight years, even in the absence of the host. In moist soil, with temperatures of 20°C to 30°C, the juveniles hatch and, if they find the root of a host plant, they penetrate and the cycle is completed in around four weeks. The transit of machinery, equipment and vehicles, carrying particles of contaminated soil, is a dispersal agent for the nematode. It can also be spread by rain, animals and seeds containing soil particles (EMBRAPA, 2023).

4.1.3 CONTROL:

Infestation should be prevented by cleaning machinery, implements, tools and footwear and by using processed seed without soil particles. Control strategies include crop rotation with non-host species, soil management (adequate level of organic matter, balanced fertilisation, absence of compaction, among others) and the use of resistant cultivars. In Brazil, there are resistant cultivars adapted to the different growing regions (EMBRAPA, 2023).

4.2 GALL NEMATODES (MELOIDOGYNE INCOGNITA AND M. JAVANICA).

Figure 12: Soya roots with galls

Source:

4.2.1 SYMPTOMS:

In infested areas, it occurs in clusters, where the leaves of the affected plants usually show chlorotic spots or necrosis between the veins ("carijó" leaf). Sometimes there may be no reduction in the size of the plants, but at the time of flowering there is intense pod abortion and premature ripening. On the attacked roots, galls are observed in varying numbers and sizes, depending on the susceptibility of the soya cultivar and the population density of the nematode. Diagnosis requires analysing soil and/or root samples in a nematology laboratory (EMBRAPA, 2023).

4.2.2 DEVELOPMENT CONDITIONS:

The nematodes that cause galls parasitise a large number of plant species and survive on most weeds, making control difficult. The life cycle is greatly influenced by temperature. At 25°C, the cycle is completed in between three and four weeks (EMBRAPA, 2023).

4.2.3 CONTROL:

The most efficient control methods are crop rotation and the use of resistant cultivars. Crop rotation must be well planned, since most cultivated species multiply gall nematodes. For this reason, the species of Meloidogyne must be correctly identified and, if possible, the breed that occurs. When rotating crops, it is important to include green manure species in order to recover the soil's organic matter and microbial activity. Direct sowing helps to reduce the spread (EMBRAPA, 2023).

- **WHEAT (TRITICUM AESTIVUM, TRITICUM DURUM, TRITICALE SP)**

Figure 13: Wheat:

Source:

Diseases can negatively affect the development of the wheat plant. We've brought you some diseases that affect wheat in various ways, and which can cause the loss of several hectares if they're not looked after properly. (SANES, T.P).

5. YELLOW SPOT (DRECHSLERA TRITICI)

Figure 14: Wheat with yellow spot:

Source:

It is the main leaf spot found on wheat crops in southern Brazil. It has been reported as one of the diseases that most affect grain yields in several countries, such as Brazil and the United States. Yellow spot epidemics can reduce yields by more than 50% when climatic conditions are favourable to the fungus' development, especially in wheat monoculture areas and on susceptible cultivars. The fungus can produce at least three selective toxins that induce leaf necrosis (Toxin A) or chlorosis (Toxin B and Toxin C) and aid in the infection and colonisation processes. Based on the ability to produce these toxins, the fungus has been classified into 8 races. In Brazil, to date, only races 1 (ToxA and ToxC) and race 2 (ToxA) have been detected in similar frequency. There is still no scientific confirmation in Brazil of the races capable of producing ToxB (EMBRAPA, 2020).

5.1 SYMPTOMS:

The initial symptoms of yellow spot are small (2 mm) dark brown lesions that develop into elliptical lesions with a brown centre and a yellowish halo. The lesions often coalesce as they grow, resulting in large necrotic leaf areas. Depending on the cultivar's reaction, yellow spot can be It can be confused with helminthosporiosis (brown spot) or septoria (glume blotch), which have similar leaf lesions. In monoculture areas with consecutive days of rain, susceptible cultivars show high disease severity (EMBRAPA, 2020).

5.2 CONTROL:

The main control strategies consist of the use of crop rotation, associated with the use of healthy seed and/or seed treatment with specific fungicides. The use of a moderately resistant cultivar, the nutritional balance of the plant (avoiding nitrogen deficiency) and the application of fungicide to the aerial organs by monitoring the disease complement the control strategies aimed at keeping the disease below a threshold of economic damage (EMBRAPA, 2020).

6. HELMINTHOSPORIOSIS (DRECHSLERA SOROKINIANA)

Figure 15: Wheat with helminthosporiosis

Source:

It occurs sporadically and rarely causes epidemics in wheat cultivation in southern Brazil. This disease prevails in warmer wheat-growing regions when there are continuous periods of wet weather. Depending on the cultivar's reaction, increased leaf severity can reduce yields and grain quality. When the fungus colonises the ear, it can infect the grains giving rise to the symptoms of "black point on wheat grains", which can interfere with the quality of the flour (EMBRAPA, 2020).

6.1 SYMPTOMS:

On the leaves, the symptoms begin with small necrotic spots, brown in colour, elliptical to rectangular in shape, 1 to 2 mm long and with a defined dark brown to black edge. On the ears, spots form on the glumes similar to those on the leaves. Infected grains may have the tip of the shield black ("black tip" symptom) (EMBRAPA, 2020).

6.2 CONTROL:

The use of healthy seed and/or treatment of infected seed with specific fungicides prevents the introduction of the pathogen into the crop. Crop rotation with non-host species reduces and/or eliminates the inoculum present in infected crop remains. Applying specific fungicides to the aerial organs is a control strategy to keep the disease below a threshold of economic damage. The presence of volunteer plants and secondary hosts in the growing area should be avoided. When available, the use of moderately resistant cultivars is indicated in regions with a favourable environment (EMBRAPA, 2020).

7. SEPTORIOSIS

Figure 16: Wheat with septoria

Source:

It is a disease that occurs sporadically in Brazil. When reported, it predominates in no-till and monoculture areas, triticultural crops with continuous periods of wetness and mild temperatures from the spike stage onwards. Damage mainly results from a reduction in grain mass, reflecting the severity of the leaf spot and/or the darkening of the glumes (EMBRAPA, 2020).

7.1 SYMPTOMS:

The initial symptoms on the leaf begin with a slightly yellowish spot that turns brown to bronze-brown. The spots are oval to elliptical in shape, usually 1-5 x 5-15 mm. Small light brown spots can be seen in the necrotic tissue, characteristic of the fungus' fruiting body (pycnidium). The fungus can infect the nodes of the thatch, which turn a dark brown colour and may have a "speckled" appearance due to the formation of pycnidia. Infected glumes show dark brown to violet necrotic tissue, starting at the tip and usually extending halfway down, causing the symptom "glume blotch", with pycnidia visible in the necrotic tissue. As the symptoms develop, the spots increase in size and can coalesce almost the entire leaf limb, often making it difficult to identify in the field (EMBRAPA, 2020).

7.2   CONTROL:

The main control strategy is crop rotation with non-host species, which reduces and/or eliminates the inoculum present in infected crop remains. The use of healthy seed and/or treatment of infected seed with specific fungicides prevents the introduction of the pathogen into the crop. Applying specific fungicides to the aerial organs is a control strategy to keep the disease below a threshold of economic damage. The use of moderately resistant cultivars, when available, is indicated in regions with a favourable environment (EMBRAPA, 2020).

8. POWDERY MILDEW (BLUMERIA GRAMINIS)

Figure 17: Wheat with powdery mildew:

Source:

It is a foliar disease that reduces yield and grain quality in susceptible wheat cultivars in harvests with mild temperatures and low rainfall. Within the subspecies that infect wheat, there are many races that are classified by their ability to infect different wheat cultivars (EMBRAPA, 2020). The predominance of these races can change over the course of agricultural seasons and between different triticultural regions, which can alter the resistance reaction of cultivars. In addition, there is the emergence of new races that can infect cultivars previously considered resistant (EMBRAPA, 2020).

8.1 SYMPTOMS:

The characteristic symptom of powdery mildew is the appearance of isolated, greyish-white colonies of the fungus (mycelium and conidia) (powdery mildew appearance), usually on the upper part of wheat leaves. The first powdery mildew colonies can already be seen on wheat seedlings shortly after emergence under favourable weather conditions. In severe epidemics, all the green parts of the plant above ground can be colonised, such as the stalk, ear and tiller. Symptoms progress from the lower leaves to the upper ones. Colonies of the fungus can grow and join together, colonising the entire leaf. Older colonies change colour from white to grey and can form small black structures called cleistothecia. Moderate to severe infections can lead to leaf death and affect photosynthesis, tillering, grain size and number, among other things. Leaves colonised by powdery mildew may show symptoms known as "green islands" towards the end of the cycle, due to the immobilisation of nutrients in the leaf, such as nitrogen. From a distance, a wheat crop infected with powdery mildew may show a yellowish colour, similar to nutrient deficiencies or excess moisture in the soil. To ensure a correct diagnosis, the crop needs to be inspected (EMBRAPA, 2020).

8.2 CONTROL:

The use of resistant cultivars is the most efficient control measure to reduce the damage caused by powdery mildew. Since the predominant races in different regions and years can change and new races can emerge capable of infecting cultivars previously considered resistant, it is important to know the updated resistance reaction of the cultivar to be sown. As it is a fungus that does not survive in wheat straw, rotation is a good idea. of crops has no effect on control. Although it does not survive on wheat seed either, seed treatment with a systemic fungicide specific for powdery mildew can be carried out, protecting the initial leaves and delaying the intensity of the disease. Applications of specific fungicides to the aerial part based on disease monitoring are also part of powdery mildew management (EMBRAPA, 2020).

9. LEAF RUST (PUCCINIA TRITICINA)

Figure 18: Wheat with leaf rust

Source:

It is found wherever wheat is grown in the world. It has a greater impact in humid regions with heavy dew and mild temperatures (15-22°C). Leaf rust can cause significant damage to susceptible cultivars. Although there are more than 50 races of the pathogen, currently in Brazil, most cultivars are considered moderately resistant, contributing to an increase in national wheat productivity (EMBRAPA, 2020).

9.1 SYMPTOMS:

It is usually seen on the leaves, but can also be found on the glumes and aristas. Symptoms begin with small yellowish oval spots at the site of infection. As it develops, these spots develop into rounded pustules (approximately 1.5 mm in diameter), which are yellow-orange in colour and may or may not have a yellow halo. Inside the pustules, using a hand-held magnifying glass, you can see the uredospores, which are easily released from the pustules and often end up accumulating on the leaf surface, forming a "uredospore dust". Some cultivars show a plant defence response known as the hypersensitivity reaction, with the formation of regions Chlorotic spots that can cover the entire leaf. As the disease progresses, the pustules can take on a black colour (telia), and are oval in shape, covered with the epidermis (EMBRAPA, 2020). Severely infected plants have dry leaves and may still have green stalks and ears (EMBRAPA, 2020).

9.2 CONTROL:

The main control strategy is the use of resistant and moderately resistant cultivars. Applications of specific fungicides to the aerial part, based on disease monitoring, are the main control strategy for susceptible cultivars, and complementary for moderately resistant cultivars, in order to maintain the effectiveness of the resistance genes (EMBRAPA, 2020).

10. VNAC (BARLEY YELLOW DWARF VIRUS (BYDV))

Figure 19: Wheat with venac:

Source:

It is widely distributed throughout the world, affecting qualitative and quantitative aspects in economically important cereals such as wheat, barley, oats and rye. In wheat, damage of approximately 20 per cent to grain yield has normally been observed in the field. However, studies under controlled conditions have observed potential damage from the disease of up to 85 per cent, depending on the viral strain, the vector population and the cultivar. The earlier the infection, the greater the damage. In no-till areas, the constant presence of virus host plants such as oats has ensured that the virus and insect vectors are maintained throughout the year (EMBRAPA, 2020).

10.1 SYMPTOMS:

They can easily be seen on the flag leaf, which becomes erect and yellow or purplish in colour. Infected plants show reduced growth and are usually distributed in clumps. Severely infected plants have blackened glumes. A reduction in leaf mass and root mass can also be observed, as well as the number and weight of grains (EMBRAPA, 2020).

10.2 CONTROL:

Control strategies include the use of cultivars that are moderately resistant and/or tolerant to the virus, as well as the management of aphid vectors through chemical or biological control. Chemical control should be carried out through seed treatment and applications during the crop cycle (by monitoring the aphid population). Biological control of vectors includes parasitoids, microhymenoptera from the Aphidiidae and Aphilinidae families (EMBRAPA, 2020).

- **RICE (ORYZA SATIVA)**

Source:

There are various rice diseases that can be very damaging to the health of the plant if they are not treated properly and correctly.Here are some of the types of diseases that can affect this crop

11. BRUSONE (MAGNAPORTHE GRISEA)

Figure 21: Rice with Brusone

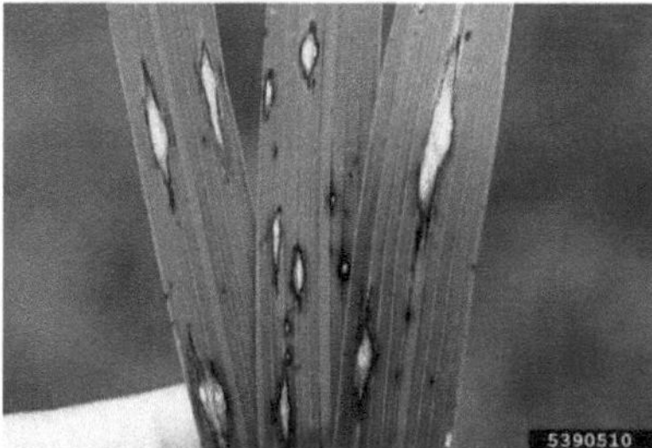

Source:

Brusone, caused by the fungus Magnaporthe oryzae (Herbert) Barr, imperfect form Pyricularia oryzae (Cooke) Sacc., is considered the most destructive rice disease and occurs throughout Brazil. Damage is variable, being greatest in upland rice, in the Centre-West region, and can compromise production by up to 100% in years of epidemic attacks. It also occurs in several common grasses in rice and wheat fields, such as Cenchrus echinatus, Eleusine indica, Digitaria sanguinalis, Brachiaria plantaginea, Echinochloa crusgalli, Rhynchelytrum roseum, Hyparrhenia rufa and Pennisetum setosum (EMBRAPA, 2024). The main sources of primary inoculum are infected seeds and crop remains. Secondary infection comes from sporulating lesions on infected leaves.All stages of the disease cycle are highly influenced by climatic factors. The deposition of dew or raindrops on the leaves is essential for the germination of conidia and the start of infection. In general, high temperatures (25 °C to 28 °C) and humidity above 90 per cent are required (EMBRAPA, 2024).

11.1SYMPTOMS:

The disease occurs from the seedling stage until the crop matures. Symptoms on the leaves begin with the formation of small brown necrotic lesions, which increase in size and become elliptical, with brown margins and a grey or whitish centre (Figure 1). Under favourable conditions, the lesions coalesce and cause the death of the leaves and often the entire plant (Figure 2). The symptoms The characteristic lesions on the nodes are brown in colour (Figure 3), which can reach the areas of the stalk close to the attacked nodes. Infection of the node at the base of the panicle is known as neck brusone (Figure 4), as a brown lesion surrounds the nodal region and causes strangulation. The grains may be completely shrivelled. The panicles turn whitish and are easily identified in the field. Various parts of the panicle, such as the rachis, primary and secondary branches and pedicels, are also infected (EMBRAPA, 2024).

11.2CONTROL:

The damage caused by brusone can be reduced by the use of resistant cultivars, cultural practices and fungicides, used in an integrated way in the management of the crop, such as: adequate soil preparation; balanced fertilisation, avoiding exaggerated vegetative growth of the plant; use of seeds of good phytosanitary and physiological quality; planting done in a minimum period of time and started in the opposite direction to the predominant wind direction; incorporation of cultural remains; uniform planting depth; recommended sowing density for the cultivar or planting system; weed control; destruction of volunteer and diseased plants; good soil levelling; changing cultivars sown every 3 or 4 years; planting at the beginning of the rainy season; using fungicides applied in seed treatment and aerial spraying. Protection against brusone on the panicle is done preventatively, by spraying with systemic fungicides: one application should be made at the end of the rubberisation period and the other up to 5% of the panicles have emerged. The recommended fungicides are listed in Pre-production Inputs. The choice can be made according to the efficiency of the fungicide, its availability on the market, registration with the Ministry of Agriculture, Livestock and Supply and cost-effectiveness (EMBRAPA, 2024).

12. GRAIN SPOT (COCHLIOBOLUS ORYZAE)

Figure 22: Rice with grain blemish

Source:

Grain spot is associated with more than one fungal or bacterial pathogen and can be considered one of the main problems in rice cultivation. As well as depreciating the appearance and reducing the quality of the grains, the disease causes a reduction in grain mass and consequently a loss in industrial yield (EMBRAPA, 2024).The main pathogens causing grain spot are the following: Bipolaris oryzae, Phoma sorghina, Alternaria padwickii, Pyricularia oryzae, Fusarium spp., Coniothyrium sp., Epicoccum sp., Rhynchosporium oryzae, Sarocladium oryzae, Nigrospora sp., Curvularia spp., Phythomyces sp. and Chaetomium sp. Bacteria that cause grain discolouration include Pseudomonas fuscovagina and Erwinia spp. The fungi predominantly associated with grain spot in irrigated rice in the state of Tocantins are B. oryzae, A. padwickii and Curvularia lunata (EMBRAPA, 2024).

12.1 SYMPTOMS

The spots appear from the moment the panicles start to emerge until they ripen. Symptoms are variable and depend on the predominant pathogen, the stage of infection and climatic conditions. It is difficult for the naked eye to identify the pathogens involved in the grain spot complex by the symptom alone (EMBRAPA, 2024). The disease is favoured by rain and high humidity during grain formation. It causes lodging by causing the panicles to come into contact with damp soil, It also contributes to increased discolouration of the grains, as well as damage caused by insects, especially bedbugs, which predispose the grains to infection by microorganisms (EMBRAPA, 2024).

12.2 CONTROL

Seed treatment with fungicides reduces initial inoculum, increases vigour and initial stand. The use of resistant cultivars, when available, and the use of systemic fungicides at the beginning of panicle emission are also effective control measures (EMBRAPA, 2024).

13. BROWN SPOT (HELMINTHOSPORIUM ORYZAE VAR)

Figure 23: Rice with brown spot

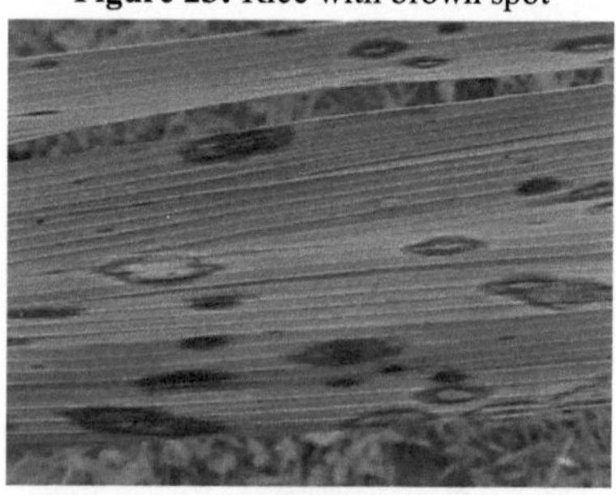

Source:

Grain spot, caused by the fungus Bipolaris oryzae (anamorph, Breda de Haan) Shoem, formerly referred to as Helminthosporium oryzae var. Breda de Haan and now considered a synonym for Cochliobolus oryzae (teleomorph), is a common disease in Brazil and is of great economic importance in rice. This fungus is one of the main pathogens that causes grain spot

(EMBRAPA, 2024).Infected seeds and crop remains are considered the initial inoculum of the disease. The dissemination of conidia through the air is responsible for secondary infection. The fungus can survive in infected seeds for 1 to 4 years, depending on storage conditions. The main factor influencing the incidence of brown spot is low soil fertility, with low levels of fertiliser, especially potassium, manganese, magnesium, silicon, iron and calcium. The optimum temperature for infection varies between 20 °C and 30 °C, relative humidity must be above 89%, and leaf wetting is favoured (EMBRAPA, 2024).

13.1 SYMPTOMS

The disease affects seedling emergence in crops sown in October, right at the start of the rainy season, and adult plants close to maturity. Infected seeds show reduced germination and, in general, spotted grains cause losses in grain yield during processing. At the seedling emergence stage, the fungus causes brown, circular or oval lesions on the coleoptile. On the leaves, the symptoms usually appear soon after flowering, with circular or oval lesions with a brown or reddish margin and a greyish or whitish centre (Figure 6). The lesions on the sheaths are similar to the typical lesions on the leaves. On the grains, the glumes have dark brown spots that often coalesce to cover the entire grain. When it manifests itself soon after the panicles are set out, the infection causes spikelet sterility (EMBRAPA, 2024).

13.2 CONTROL

Seed treatment with fungicides reduces the initial inoculum. The use of resistant cultivars, when available, and systemic fungicides at the beginning of panicle emergence are also effective control measures (EMBRAPA, 2024).

14. SCALD (RHYNCHOSPORIUM ORYZAE)

Figure 24: Scalded rice

Source:

Scald, caused by the fungus Monographella albescens (Thüm.) Parkison et al., imperfect state Rhynchosporium oryzae, is among the main rice diseases. Epidemics of the disease are common in crops planted in rotation with soya. In general, it is an important disease in environments with high rainfall. The primary sources of inoculum are infected seeds and crop remains. Climatic conditions favourable to the development of leaf scald are high rainfall, average temperature between 24 °C and 28 °C, prolonged periods of dew, high sowing density and excessive nitrogen fertilisation. Insect damage is an entry point for the pathogen.

14.1 SYMPTOMS

Leaf scald has more than one type of symptom. The most characteristic symptom is identified at the apical ends of older leaves or at the edges of leaf blades, where olive-green spots appear without well-defined margins. The lesions develop into successions of concentric bands, alternating light and dark brown colours. The lesions coalesce and cause necrosis and death of the leaf area. A severe incidence of scald, while causing a loss of leaf area, paralyses the growth of plants in the middle of the germination stage and affects the quantity and quality of the grains.

Normally, affected crops show generalised yellowing, with dry leaf tips and irregular plant height. Another type of symptom is characterised by brown spots along the leaves, similar to the initial symptoms of grain spot. Scald can also affect the sheaths and cause symptoms similar to those on the leaves. On the grains, the fungus causes small pinhead-sized spots and, in severe cases, causes discolouration of the glumes, turning them reddish-brown.

14.2 CONTROL

Seeds of good phytosanitary quality should be used. The application of fungicides in seed treatment and spraying is indicated for disease control. Crop rotation helps to reduce the incidence of the disease.

15. NARROW SPOT (COCHLIOBOLUS ORYZAE)

Figure 25: Rice with a narrow spot

Source:

Star spot, caused by Sphaerulina oryzina K. Hara (Cercospora oryzae Miyake; syn. C. janseana (Racib) O. Const.), is a problem when it occurs in high severity on the upper leaves shortly after the panicles have emerged, as it can reduce the photosynthesising leaf area, cause a reduction in grain mass and rapid ripening, as well as reducing the final yield. In susceptible cultivars, the disease can cause premature senescence and jeopardise productivity and grain quality. In Brazilian conditions, the disease is of little importance because it usually occurs at the end of the crop cycle.

15.1 SYMPTOMS

Typical symptoms are narrow, thin, necrotic spots, elongated in the direction of the veins, reddish-brown in colour, which appear most frequently on the leaves (Figure 9). Similar symptoms can occur on sheaths, stalks, pedicels and glumes. The disease is favoured by high humidity and high temperatures (28 °C). The fungus survives in crop remains and the conidia are spread by the wind.

15.2 CONTROL

The use of resistant varieties is the best way to avoid or reduce losses. Chemical control with fungicides registered for the crop and the disease is also recommended, as well as the use of healthy or treated seeds.

REFERENCES

CAMPOS, Marcelo Furian. Monitoring the soya crop in north-west Rio Grande do Sul: 2019/2020 harvest. 2021.

SOARES, R.M.; GODOY, C.V.; et al.; Manual de identificação de doenças de soja, 6th edition, Embrapa Soja, Londrina, 2023.

LAU, D.; SBALCHEIRO, C. C.; MARTINS, F. C.; SANTANA, F. M.; MACIEL, J. L. N.; FERNANDES, J. M. C.; COSTAMILAN, L. M.; LIMA, M. I. P. M.; KUHNEM, P.; CASA, R. T.; Main wheat diseases in southern Brazil: diagnosis and management. Embrapa Wheat, leaflets, 2020.

LOBO, V.L.S.; FILIPPI, M.C.C.; PRABHU, A.S.; Manejo de doenças, Embrapa, 2024.

CHAPTER 6

HARVESTING

Andrei Pigatto Leandro Cavalheiro
Joice Fernanda Lübke Bonow Andréa Bicca Noguez Martins

INTRODUCTION

Harvesting is a determining process in agriculture and plays a crucial role in defining product quality and production efficiency. The post-harvesting of agricultural produce is a series of processes carried out with the aim of preserving product quality and increasing the product's shelf life. According to MARTINS et al. (1999), grains should be harvested at the point of physiological ripeness, when they have high levels of starch, protein and water. By removing moisture through drying and proper storage, it is possible to preserve agricultural products. Drying ensures better grain quality by making it possible to harvest earlier, avoiding damage in the field due to weather conditions, insect attacks and microorganisms.

Mechanised harvesting is a fundamental tool in the production process of major crops, but when carried out incorrectly, it leads to considerable losses, reducing productivity and profits for producers.

In Brazil's wheat-growing regions, rain is common during the harvest period, resulting in a high incidence of sprouted grains, classified as inferior quality for the bread industry, with a depreciation in value of around 50%. Considering that around 20% of the wheat crop each year is sprouted grain, this would result in approximately 437,000 tonnes of sprouted wheat (BRUM, 2000).

In soya harvesting, Carvalho Filho et al. (2005) observed that the length of time harvesters are used interferes with losses, making it necessary to constantly monitor the equipment and keep it in good working order.

There is also the uniformisation of crops through desiccation with the application of herbicides to bring forward the harvest with a view to sowing the successor crop. Pre-harvest desiccation is a practice adopted to promote better conditions in the operation and enable subsequent cultivation at suitable times (FIPKE et al, 2019).

In recent years, the introduction of advanced technologies such as precision agriculture, crop forecasting applications, the use of VANTS (unmanned aerial vehicles) also popularly known as drones has transformed significantly in this sector, providing gains in terms of productivity and sustainability.

MATERIAL AND METHODS

The chapter was written following bibliographical research into the harvesting techniques used in rice, soya and wheat crops and the new technologies that are being developed and applied to improve performance in agricultural processes. Scientific articles and research published in recent years were used as a research base.

RESULTS AND DISCUSSION

For mechanised harvesting of soya beans, it is desirable for the moisture content of the grains to be between 12 and 14%. Grains with a moisture content of more than 14% may be subject to a higher incidence of latent mechanical damage during the harvesting process, while moisture contents of less than 12% may result in greater sensitivity to grain breakage and damage to the physiological quality of the seeds (França Neto et al., 2007). According to Campos (2005), the greatest yield losses occur in environments with a high frequency of rainfall, high temperatures and high global solar radiation.

The ideal time to harvest rice is basically determined by the appearance of the panicle, the length of the crop's development stages and the moisture content of the grains. It is accepted that careful visual observation allows the most favourable time for harvesting to be determined fairly accurately. The highest yield is obtained when approximately the top two thirds of the rachis are yellowed and the panicle is curved. The appropriate seed moisture content for harvesting rice is between 18% and 27% (FRANCO et al, 2013).

According to Portella et al (2011), it is recommended to harvest wheat when the grains have a moisture content of between 16% and 18%, as this is where the harvester performs best, due to lower losses caused by the threshing of the ears on the platform, and less crushing of the straw.

Precision agriculture (PA) offers a multitude of potential benefits in terms of profitability, productivity, sustainability, crop quality, environmental protection, quality of life, food security and rural economic development. Remote sensing techniques, where crop information is obtained non-destructively, quickly and sometimes remotely, have become of fundamental importance in obtaining and processing field data (BERNARDI et al, 2014).

According to Coelho (2013) the tools currently available that enable the application of PA technologies are: Global Positioning System (GPS), Geographic Information System (GIS), crop monitors and mapping, targeted soil sampling and mapping of the physical and chemical attributes of soils, sowing and application of correctives and fertilisers at variable rates, remote sensing and satellite guidance systems (GPS guidance system).

According to Souza, (2013) the use of technology to collect data in the production unit, such as mobile software, eliminates the manual form of data collection, as well as ensuring greater agility and reliability in the information collected. The tool can be installed on portable devices such as mobile phones and tablets, which contain the Android operating system. According to the same author, there is no need to invest in data collectors, which are expensive and difficult to replace, as they can only be bought in specialised shops. A device with the Android operating system can be bought easily in a number of retail shops at very affordable prices.

In a study by Melo et al (2021) on grain counting sensors used to make it possible to verify the efficiency of each part of the crop, by comparing the georeferenced map (GPS) and another map produced f r o m productivity, it is possible to verify the contribution of each part of the crop at the end of production, since the sensor provides verification o f production variability, indicating the areas with the highest and lowest production potential, which makes it possible for the producer to manage efficiently when making decisions for the next crop.

Productivity measurement is an automatic process. It is carried out using sensors installed in harvesters and capable of defining, with relative accuracy, the amount of product being harvested and the area where it was produced. (COELHO,2013) According to Alarcão (2023) the use of drones in agriculture offers significant advantages such as: monitoring, detailed and accurate images; and crop management as they have the ability to fly over large areas in reduced time and obtain detailed and accurate images, drones allow farmers a comprehensive view of the state of crops, identify problems such as diseases, pests, water and nutrient stress and provide information for harvest planning.

FINAL CONSIDERATIONS

The modernisation of harvesting through the incorporation of technologies is an indispensable factor in meeting the current challenges facing agriculture. Technological innovations not only increase the efficiency of harvesting, but also guarantee a better quality of the harvested products, promoting the sustainability of the agricultural sector.

BIBLIOGRAPHICAL REFERENCES

ALARCÃO JÚNIOR, J. C. et al. The use of drones in agriculture 4.0. Brazilian Journal of Science, v. 3, n. 1, p. 1-13, 2023.

BRUM, P.A.R. Wheat in poultry feed. Avicultura Industrial, Porto Feliz, p.14- 16, 2000

BERNARDI, A.C. De C. et al. Precision agriculture: results of a new approach.

Cnptia.embrapa.br. 2014.
CAMPOS, M. A.O et al. Losses in mechanised soybean harvesting in the state of Minas Gerais. Engenharia Agrícola, v. 25, p. 207-213, 2005.

CARVALHO FILHO, A. et al. Losses in mechanised soya harvesting in the triângulo mineiro. Revista Nucleus, Ituverava, v. 3, p. 57 - 60,2005.

COELHO, A. M. et al. Precision agriculture maps grain harvesting. infoteca.cnptia.embrapa.br. 2013.

DE MELO, L. V. et al. Analysing the advantages of using remote sensing to support precision agriculture. EnPE, v. 8, n. 1, 2021.

FRANÇA NETO, J. de B. et al. High quality soya bean seed production technology. Londrina: Embrapa Soja, 2007. 12p. (Technical circular, 40).

FRANCO, D. F. et al. Harvesting, drying, processing and seed treatment of irrigated rice. 2013.

FIPKE, G. M. et al. Productivity and economic viability of anticipating the wheat harvest by applying herbicides. 2019.

MARTINS, R.R. et al. Grain drying technology. Passo Fundo: Embrapa Trigo/EMATER/RS, 1999. 90p. (Embrapa Trigo. Documentos, 8)

PORTELLA, J. A. et al. Wheat harvest technology. In:PIRES, J. L. F.; VARGAS, L.; CUNHA, G. R. da (Ed.). Wheat in Brazil: bases for competitive and sustainable production. Passo Fundo: Embrapa Trigo, 2011. Chap. 13, p. 325-348.

SOUSA, E. et al. Optimising the coffee harvest management process using automated tools. CBPC (39.:2013: Poços de Caldas, MG) - Proceedings 2013.

I want morebooks!

Buy your books fast and straightforward online - at one of world's fastest growing online book stores! Environmentally sound due to Print-on-Demand technologies.

Buy your books online at
www.morebooks.shop

Kaufen Sie Ihre Bücher schnell und unkompliziert online – auf einer der am schnellsten wachsenden Buchhandelsplattformen weltweit! Dank Print-On-Demand umwelt- und ressourcenschonend produziert.

Bücher schneller online kaufen
www.morebooks.shop

Printed by Books on Demand GmbH, Norderstedt / Germany